China's Rise as a Supercomputing Power

Gong Shenghui & Zeng Fanjie

图书在版编目（CIP）数据

超算之路：英文 / 龚盛辉，曾凡解著；悦享怡然文化译 .
-- 北京：五洲传播出版社，2019.1
ISBN 978-7-5085-4002-3

Ⅰ . ①超… Ⅱ . ①龚… ②曾… ③悦… Ⅲ . ①报告文学－中国－当代－英文 Ⅳ . ① I25

中国版本图书馆 CIP 数据核字 (2018) 第 188756 号

作　　者：龚盛辉 曾凡解
译　　者：悦享怡然文化
出 版 人：荆孝敏
责任编辑：姜珊
助理编辑：宋歌
装帧设计：北京牧涵文化传媒有限公司

超算之路（英文）

出版发行：五洲传播出版社
地　　址：北京市海淀区北三环中路 31 号生产力大楼 B 座 6 层
邮　　编：100088
电　　话：010-82005927，82007837
网　　址：www.cicc.org.cn, www.thatsbook.com
印　　刷：北京画中画印刷有限公司
版　　次：2019 年 1 月第 1 版第 1 次印刷
开　　本：710 × 1000　　1/16
印　　张：27
定　　价：128.00 元

Preface

Supercomputing: The Most Advanced Technology

As human knowledge keeps expanding and deepening, especially with the appearance of big science, big engineering, and big data in the modern age, supercomputing has gradually become as important as scientific theories and experiments. As one of the three pillars supporting the mansion of modern science and technology, supercomputing represents China's competitiveness in science and technology.

Nowadays, supercomputing is indispensable to the development of a nation, from strategic studies on the topic of national security to improvements in people's living conditions. Supercomputing has already penetrated into people's daily life, including their work, housing, clothing, food, transportation and entertainment. People enjoy the merits of supercomputing at all times. For example, supercomputers are being used in food production. Researchers utilize supercomputing technologies to help with genetic engineering, of which the main research objects are rice, corn and pigs. The purpose of such studies is to increase food production and make food more palatable and nutrition-rich. At the same time, this research can help pigs grow faster and produce better-quality meat, which better benefits human health. In addition, with supercomputing technologies, the cycle of new drug development has been shortened from a few years or even over a decade to less than a year or even a few months; the genetic testing process for cancer patients has been reduced from a month or two to a few minutes. In this way, patients can receive timely treatments. Weather forecasting is yet another field in which supercomputing plays a role. It only takes a few minutes for a supercomputer to calculate the weather conditions in the coming week. Within 24 hours, a supercomputer can finish computations that would have taken a few years or even decades in the past. Moreover, forecasting earthquakes, tsunamis and other natural disasters has also become possible with the help of supercomputing technologies. The entertainment industry also depends on supercomputing to produce fantastic animation rendering in many movies and TV series that audiences really enjoy watching.

At present, there are challenging problems that rely on supercomputers to solve. Such problems occur in the manufacturing of automobiles, airplanes and ships, in terms of how to improve aerodynamic and hydrodynamic structures to reduce fuel consumption and noises while enhancing collision avoidance system performance and riding comfort. Supercomputers are also in use to determine how to prevent and mitigate the impact of climate change, how to help patients with more effective and revolutionary medical treatments and how to reduce casualties and losses caused by earthquake by identifying what kind of measures we need to take to develop our capacity for prediction and preparedness. In addition, supercomputing plays a crucial role in modelling and experimenting with the movement of celestial bodies, simulating incidents that influence social well-being and safety while finding corresponding measures and plans, exploring materials with high economic value and discovering the pattern of human activities and social development. These issues cover almost every field of scientific research and they are closely related to all sectors of social life. A great range of subjects are involved, including vehicle manufacturing, weather forecasting, bioinformatics, earthquake monitoring, earth science, astrophysics, public health, materials science, and human/

organizational system studies.

In other words, it is extremely difficult for human beings to explore advanced science these days without supercomputing technologies. Supercomputers really deserve to be called "a nation's treasure!"

On 14 February 1946, on the campus of the University of Pennsylvania in the United States, Professor Mauchly, Lecturer Eckert and the founder of modern theoretical computer science, Von Neumann, raised glasses of champagne to celebrate the successful creation of the first electric numerical computer in the world. At the same time, however, China was still on the eve of an all-out civil war. It took years for China to start its computer engineering programmes. As a result, China, as a late starter in developing this emerging technology that would later exert great impacts on the people's lives, was far behind its counterparts.

However, there is a great team that keeps the motherland in mind. With great ambition, strong cooperation and tenacious effort, this team, Yinhe, has been tackling the problems in computer technology. Despite great impediments and challenges from developed countries, Yinhe has tried its best to catch up and successfully developed China's first special-purpose computer, the first general-purpose transistor computer, the first computer with a speed of one megaflops, the first supercomputer with a speed of 100 megaflops, the first parallel supercomputer with a speed of one gigaflops, the first super-large-scale parallel supercomputer with 10 gigaflops and so on. The creation of many scientific miracles such as "China Core," "China Kylin" and "China's First Website" has made Yinhe the best supercomputing team in China.

In the early 21st century, as advanced supercomputing technology of 100 teraflops has been successfully achieved, human beings are facing a series of great difficulties to advance key supercomputing technologies. This means that countries in the world are standing at the exact same starting line for developing a new generation of supercomputers.

Yinhe has seized this historic opportunity, taking the lead in breaking through the mainstream technology of the new generation of supercomputers in the form of the CPU+GPU heterogeneous parallel system.

In 2010, the second stage of the "Tianhe-1" computer system claimed the top spot on the list of the world's supercomputers.

The "Tianhe-2" topped the list of the world's supercomputers five times since its birth in 2013, which demonstrates China's leading role in the development of supercomputing technologies.

Contents

Contents

Contents

Chapter I
The Imagination of Youth

On 17 November 2010, an upgraded version of the "Tianhe-1" system, developed by the Institute of Computing Technology under the National University of Defense Technology, achieved glory by taking the number one position on the list of the world's TOP500 supercomputers.

On 17 June 2013, at the 41st International Supercomputing Conference held in Leipzig of Germany, the "Tianhe-2" system restored its former glory by ranking as the world's number one supercomputer. Since then, the "Tianhe-2" system has retained top placement on five consecutive TOP500 lists.

I. The "Tianhe-1" Attracted Worldwide Attention

At 5:30 p.m. on 16 November 2010, the award ceremony of the TOP500 took place in New Orleans, Louisiana, in America. At the focal point of the audience and cameras, Hans Meuer, a famous computer expert, professor at the University of Mannheim, and founder of the international TOP500 list, stepped steadily onto the rostrum and announced the top three supercomputers on the TOP500 list: the "Tianhe-1" system developed by the National University of Defense Technology, the Jaguar system at the Oak Ridge National Laboratory of America, and the Dawning Nebulae developed by China's Dawning Information Industry Corporation.

In situ testing by experts from the TOP500 project found that the "Tianhe-1" system achieved a peak computing rate of 4.7 petaflops and a sustaining computation rate of 2.566 petaflops. Its work within an hour equalled 340 years of work by all the 1.3 billion Chinese people, and its work in a day equalled 620 years of work by an advanced dual-core desktop computer. The total memory size of the system was 1,000 trillion Chinese characters, the equivalent of a huge library with one billion books, with one million characters in each book.

The peak computing rate of the "Tianhe-1" system was over twice that of the Jaguar system, which ranked in the second place.

Experts from all over the world commented that it was undoubtedly the most dramatic dark horse in the history of computer technology in the world.

Arthur Trew, the Director of the Edinburgh Parallel Computing Centre under the University of Edinburgh, told reporters, "This is an interesting change. For many years, America has always been proud of having the world's fastest supercomputers. But now, the honour belongs to China."

A computer expert from the Virginia Polytechnic Institute in America said that it signified the faltering of American dominance in this technological field, and it might even shock the economic prospects of America.

Jean Genuoer, Director of the Digital and Analogue Information Project in the French Atomic Energy Commission, thought that the significance of the "Tianhe-1"'s computing rate reaching world record levels extended far beyond the computer itself. It signified a giant leap forward for China's technological capabilities, as well as China's growing economic competitiveness.

Associate Professor Mitsuyoshi Hiratsuka from the Tokyo University of Science in Japan thought the "Tianhe-1" showed that China could develop the most critical technologies in the field of electronics.

The German weekly magazine Der Spiegel commented that the West often labelled technologies developed by China as "copies" of others, but the "Tianhe-1" system indicated that China had already become an innovative country.

...

There was no doubting that international competition in this field was extremely fierce. Only eight months later, Japan's K supercomputer surpassed the "Tianhe-1" system. Soon, the Sequoia and Titan supercomputers developed by America took the number one ranking in succession, while the "Tianhe-1" system was ranked in eighth place.

While the world was amazed by the rapid development of supercomputers, the researchers of the Tianhe system were surprisingly calm. For them, being surpassed only meant the beginning of a new round of competition. The Tianhe staff knew very well that international competition in the supercomputing field was like a sprint, with a starting point but no finishing point. There was no permanent champion, only a cycle of constant surpassing and being surpassed.

As Yang Xuejun, the chief designer of the "Tianhe-1", said, "Since the birthday of the 'Tianhe-1', the research on the 'Tianhe-2' system has been started. In the fiercely contested international supercomputing field, only by striving to surpass others can one keep one's leading position."

In June 2013, the "Tianhe-2" system restored its former glory by grabbing the TOP500 crown at the 41st International Supercomputing Conference in Leipzig, Germany.

The "Tianhe-2" system consisted of 170 cabinets. Its work within an hour equalled 1,000 years of work by 1.3 billion people with calculators, and its total memory size was the equivalent of a library with 60 billion books, with 100,000 characters in each book.

It has created a new "Chinese speed" with genuine high performance and high efficiency.

Jack Dongarra, an expert from the international TOP500 project and a professor at the University of Tennessee, said, "The size of the 'Tianhe-2' system is almost the same as that of the American Titan supercomputer, but it is twice as fast as the Titan, which is very stunning."

Professor Sebastian Schmidt from the Julich Research Center in Germany said, "The 'Tianhe-2' system is one of the best computers in the world. Due to its outstanding performance, I believe that it is able to solve many problems in the scientific field."

Up to now, the "Tianhe-2" system has retained the first place on the TOP500 list for the fifth consecutive time. From the "Tianhe-1" system, which was considered to be a dark horse in the international supercomputing field, to the "Tianhe-2" system, which won five consecutive championships, China's supercomputing dream has finally come true.

On the way to the peak were the winding but solid footprints of many Chinese people working in the computer industry, who had contested with international computer powers for more than half a century.

II. The Invigorators of the "Eastern Lion"

Napoleon Bonaparte once said, "China is a sleeping lion."

On 1 October 1949, the Eastern lion woke up, as Mao Zedong solemnly declared, "The Central People's Government of the People's Republic of China has been established." Although the awoken Eastern lion had a huge skeleton, it had no strong muscles and seemed to be a bit thin and weak. How could they make the Eastern lion strong as soon as possible? Leaders of the PRC cast their thoughtful eyes to the whole world and plumbed the depths of history.

No matter whether it was ancient China, or early modern Britain, or modern America, a country's long-term stability as a world power is based on political and military revolutions in the early stages and major scientific and technological developments in the following stages. Science and technology were the most effective "invigorators" of the Eastern lion, making it strong as quickly as possible.

Five days later, Zhou Enlai appealed at the second session of the 2nd Chinese People's Political Consultative Conference, "March toward modern science and technology!" Zhou directed the State Planning Commission, the Chinese Academy of Sciences and other relevant departments to work together to formulate the 12-Year Program for the Development of Science and Technology (1956–1967). On 25 January 1956, Mao Zedong proposed at the supreme state conference that "The Chinese people shall have a long-term and ambitious plan to improve China's backward situation in the field of economy, science, and culture within several decades, and quickly reach the advanced level in the world."

He said, "This long-term plan is aimed at, based on our needs and all possibilities, introducing the most advanced international scientific achievements to China as soon as possible, reinforcing the most urgently needed scientific departments in China as soon as possible, planning our scientific research on the basis of existing scientific achievements in the world, and promoting the most urgently needed scientific departments in China close to the advanced level in the world by the end of the third five-year plan."

In October 1956, Nie Rongzhen was appointed as the Vice Premier of the State Council by the central government and took charge of China's scientific and technological development. His first mission after taking office was to lead the formulation of a 12-year plan for the development of science and technology.

In the plan, electronics was listed as the third item of 12 major national scientific and technological missions, among which computer technology was one of the most urgent.

As required by the plan, in June 1956, China established the Preparatory Committee for the Institute of Computing Technology, whose director was Hua Luogeng, a famous Chinese mathematician and Director of the Mathematical Institute of the Chinese Academy of Sciences. At the end of 1956, China sent a group of 20 people to visit the Academy of Sciences of the former Soviet Union and learn its computer technologies. Shortly afterwards, China ordered the drawings of two types of computers and commenced to reproduce electronic computers. China had finally embarked on the way to explore computer technologies.

III. The Courage of Youth

In April 1958, the Naval Engineering Department of the PLA Military Institute of Engineering set up the 901 (331, in the beginning) electronic numerical computer research team, which was under the direct leadership of Deputy Dean Ci Yungui. Liu Kejun and Hu Shouren were appointed as the head and political commissar of the team, respectively, and its members included Hu Keqiang, Chen Fujie, Lu Jingyou, Geng Huimin, Zhang Maya, and Sheng Jianguo.

This was a group young scientific researchers, with an average age of 25. Ci Yungui, the deputy dean, directly led this mission and was the eldest in the team at just 40 years old; the political commissar Hu Shouren was just over 30 and the head of the team, Liu Kejun was only 25.

After the establishment of the People's Republic of China, in order to meet the country's needs, Ci Yungui, then a lecturer at Tsinghua University, entered the service. In April 1950, he was appointed Associate Professor at Dalian Higher Naval School. In 1955, he was transferred to be the head of the radar section of the Naval Engineering Department at the PLA Military Institute of Engineering in Harbin, and became the deputy dean of the department the following March. In July 1957, he found an article in the Journal of American Institute of Radio Engineers which comprehensively introduced electronic computers. He read the article again and again, as if he had received a valued treasure. Inspired by the article, he developed a strong urge to contribute to the country's computer industry.

Liu Kejun pursued his postgraduate study at the Harbin Institute of Technology after his graduation from Tsinghua University in 1954. In 1956, after graduating from the Institute, he worked in the PLA Military Institute of Engineering. In 1956, Liu learned from a journal that Britain was developing computers. He was instantly fascinated by this "mysterious stuff". Although not even knowing the shape of a computer, Liu referred to a popular science article and made use of the knowledge about mechatronics and automation he had gained in his undergraduate and postgraduate study, and set free his rich imagination and deep thought about the world of computer science. He pondered: what is the difference between an abacus and a computer? What is the connection between abacus formulae and computer software? How do computer components correspond to the abacus beads? How does the computer carry out shift operations? How are the data input into the machine? After months of pondering, he finally figured out the basic principles of computers and came up with the idea to produce torpedo boat directors with computer technologies.

In the summer of 1957, as proposed by Senior General Chen Geng, China's Central Military Commission decided at its 106th meeting to send a delegation of the PLA Military Institute of Engineering, headed by Liu Juying, the vice president of the Institute, to visit the former Soviet Union, Poland, and Czechoslovakia. Ci Yungui, then serving as the Deputy Dean of the Naval Engineering Department, was a member of the delegation. Liu Kejun, then a lecturer of electrical automatic command, also joined the delegation as an interpreter and secretary. In the former Soviet Union, Liu finally saw the coveted computer. This "mysterious stuff" was like a huge magnetic field, which firmly grasped his eyes and held him fast.

After returning to his institute, the first thing for Liu was to submit "The Report about Developing Shipboard Electronic Computers and High-speed Electronic Directors for Fast Boats" to the leaders of his department. He wrote in this report, "At the end of 1956, based on the development of international electronic computer technologies, I came up with the idea to solve problems of the torpedo boats with electronic numerical computers. After continual exploration, I recommend developing shipboard electronic computers for the navy."

The electronic computer was a brand-new technology. The young PLA Military Institute of Engineering assigned this "young" mission to a young team.

Firstly, most of the research team had never seen a computer. They knew nothing about computers, not even the binary operation of computers.

Secondly, they had no laboratory, no equipment, and no raw materials.

Apart from a Russian and an English popular science article about the basics of computers, these young people had nothing.

IV. Borrowing Chickens to Lay Eggs

One day in early 1958, Kang Jichang, a member of the PLA Military Institute of Engineering, stepped up to a podium, opened a computer science pamphlet written by a member of the former Soviet Union Academy of Science, the academician S. A. Lebedev, and began to introduce the ABCs of electronic computers to members of the development team. The work on China's first self-designed and developed "901"-type electronic special-purpose computer officially started.

The war was a veritable "cross the river by feeling the stones". Among the researchers, only Ci Yungui and Liu Kejun had some knowledge about computers, while others were almost completely clueless about computer knowledge. They could only learn from capable colleagues, and then discuss and practice while learning. In their own words, "with books in one hand and a multi-meter in the other, we were groping our way forward." Fortunately, most of the members came from the Radar Teaching and Research Office, which was close in subject matter to electronic computing since they belong to the same "field" (electronics). Members were familiar with the basic theories, so it was easy for them to draw inferences and comprehend by analogy.

Members quickly reached the first major milestone—they mastered the basic knowledge of electronic computers. Based on this knowledge, they continued ahead and initiated and discussed the preliminary development idea of the "901" computer. The third milestone was discussing and determining the development plans, steps, methods and the schedule.

However, their ardour, courage and faith alone for the scientific research were far from enough. On the battlefield, with a spirit of not being afraid of sacrifice and an overwhelming impetus, picking up one's gun and lowering one's head may be enough for success. However, on the battlefield of scientific research, in addition to spirit and courage, certain material conditions are also needed. Scientific research is similar to chickens laying eggs. If one wants eggs, one must first have chickens, because it is not possible to get something from nothing. The development team was desperate for the egg, but they did not have the chicken—they did not have any laboratories, scientific equipment or raw materials.

What could they do? There was only one way, which was to borrow chickens to lay the eggs.

They first borrowed the radar laboratory from the department to open up the first "base area" for this crucial war, and then they split into two groups to solve respective problems.

Ci Yungui led a group of people and went to Beijing to ask for support and help from superior leaders and institutions and to learn from the Institute of Computing Technology of CAS in order to obtain the equipment for the research. The core mission was developing magnetic core memory. The stored-program computer proposed by Von Neumann was a major milestone. It gave the first human computer, ENIAC, truly scientific computing capabilities and became a turning point in the history of world computer technology. The memory was the key technology of developing the stored-program computer, which meant the development of memory was even more difficult.

Another group of people stayed in Harbin and were responsible for basic circuit design experiments, as well as the logic design of the arithmetic logic unit (ALU), control units and the development of other components.

The Beijing trip of Ci Yungui and his group won great support from the superior institutions and associate organizations. At that time, the Institute of Computing Technology of the CAS was developing 103 minicomputers and 104 mainframes with the help of the former Soviet Union. The Institute of Computing Technology arranged people from the PLA Military Institute of Engineering into a temporary research team and gave them a laboratory like other research teams in the Institute and provided experimental equipment and tools for them. They could borrow various materials, get various experimental equipment and participate in various activities, including academic activities. Their accommodation was also arranged.

The selfless help of the Institute of Computing Technology made them feel warm and encouraged while they also realized the heavy and urgent tasks on their shoulders. They devoted themselves to the development work for about seventeen or eighteen hours every day. The laboratory, canteen and dormitory were the only destinations in their daily life.

With the non-stop work, the development proceeded rapidly. The circuit was finalized, the magnetic core test board was assembled and the development of magnetic core memory started.

However, as the research and development was drawing to a close, another problem was becoming increasingly urgent—the raw materials for the complete machine. Electronic components were rare in China and the PLA Military Institute of Engineering could not solve this problem for the team.

Ci Yungui could do nothing but ask for help again from the superior and associate organizations.

He went to the Ministry of the Electronic Industry. The leader said, "It is a significant project related to the coastal defence security and must be fully supported!"

He went to the Department of Communications of the PLA Navy for assistance. The ministry leaders said, "Your difficulties are ours and we can solve them together!"

In this way, when people in Beijing returned to the PLA Military Institute of Engineering in early August of 1959, they not only brought back a large amount of materials, research and development tools, a core tester for key test equipment and key components of the magnetic core memory, but also brought back the components to assemble the complete machine.

Meanwhile, the people who stayed in Harbin also conquered a series of difficulties, such as the basic circuit design and experiment and the logic design of the arithmetic logic unit and the control unit. The first phase task of the research was completed. They spent the days in diligent and industrious work. The newspaper Engineering reported in May 1958 that "Liu Kejun, Hu Shouren and other comrades finally completed the design of a new type electronic computer after a month of tenacious work. During the design process, they worked about 12 hours every day and even brought their lunch and dinner to the laboratory. Even on Saturdays and Sundays, they were still digging into work in the laboratory."

After the two groups gathered in the PLA Military Institute of Engineering, the final attack—the complete machine assembly and the test—started.

Everyone was feeling their way forward, so there were inevitable difficulties. Once difficulties emerged, the development team would hold a meeting to gather plans to confront the difficulties, moving consistently towards the target.

One day, another problem occurred that was a significant barrier to further work, and one which could not be overcome in a short time: the magnetic core memory was continuously unstable. The structure of the core stack was complicated and it was very difficult to check and repair. The National Holiday was coming and everyone was nervous.

The political commissar of the department, Deng Yifei, heard the news and came. After hearing the report of the development team and looking at the bloodshot eyes of the comrades, he decisively ordered everyone to immediately go home to sleep and clear their minds. They should analyse the memory and replace the faulty component, establish the repair team immediately and start work again tomorrow. He told them not to give up until the problem was solved.

With the determination to fight to win, after continuous work for three days and nights, the repair team finally found the faulty magnetic core and replaced it successfully. After the tests, the magnetic core memory was finally stable and achieved the design objective.

On 28 September, early in the morning, the machine started to calculate. It was the last battle of the "901" war. All the related leaders from the administration and department came, and the operator tapped the keyboard deftly under the gaze of dozens of eyes, inputting an elliptic integral question into the machine. A moment later, the printer was squeaking.

Ci Yungui carefully compared the printed paper with the answer and announced loudly, "The calculation result is totally correct!"

The first vacuum tube special-purpose computer designed and developed by Chinese engineers was finally born.

Everyone cheered and embraced each other, with tears in their eyes.

On 28 September, the "901" ran normally. On the 29th, the various operations performed by the machine were all correct.

After the appearance of the world's first electronic numerical computer, ENIAC, the United Kingdom and the former Soviet Union imitated the device and started their own computer development. The United Kingdom spent two years to develop the first numerical computer, known as the Manchester computer. From the establishment of the project to the completion of the assembly, the former Soviet Union took five to six years on its first electronic numerical computer, NORC.

The first Chinese electronic numerical minicomputer, the "103" computer, from the moment of importing drawings from the former Soviet Union to beginning the trial-manufacture to successfully calculating a question for the first time, only took several months.

The first mainframe machine, the "104" computer, from the moment of importing drawings from the former Soviet Union to beginning the trial-manufacture to successfully calculating a question for the first time, also only took several months.

To design and develop the first Chinese electronic numerical special-purpose computer prototype, the PLA Military Institute of Engineering took less than a year from the establishment to the successful calculation of the machine!

It is the genuine "Chinese Speed"!

V. The Prospective Vision

In 1958, China's central military commission decided to hold an exhibition in Beijing reflecting the results of the technological revolution in Chinese universities. Director Chen Geng instructed the Institute to select a group of achievements from hundreds of scientific research achievements to participate the exhibition.

During the exhibition, Chen Geng ignored his doctor's advice and went to the exhibition hall to watch the exhibits several times. He listened to the introductions and warned presenters repeatedly not to exaggerate. In those days, Chen Geng was cheerful all day.

After the exhibition, Chen Geng called the leaders of the Central and Military Commission one by one, saying, "the PLA Military

Institute of Engineering is reporting to the Central Committee; come and have a look at all these high-level achievements in scientific research." The leaders readily agreed after hearing the call.

On the opening date of the exhibition, Chen Geng waited at the gate of the exhibition hall in the early morning. Soon after, Liu Shaoqi first came, followed by Zhu De, Deng Xiaoping, Peng Dehuai, Ye Jianying, Su Yu, Huang Kecheng, Lin Boqu and other national and military leaders. For a time, the exhibition hall was crowded with generals and leaders.

Chen Geng hurriedly apologized, "I am so sorry that the hall is so small and crowded."

Peng Dehuai joked to him, "As long as the PLA Military Institute of Engineering has more products, we are very happy to huddle together here."

After the "901" computer and other scientific achievements were moved to the Beijing Navy Community for further exhibition, Zhou Enlai, Chen Yun, Lin Biao, Chen Yi, Luo Rongzhen, Xu Xiangqian and other leaders also came to visit.

Zhou Enlai was very interested in the "901" computer, observing and asking questions very carefully. He said affectionately to Chen Geng and Ci Yungui, "The achievement is very remarkable and we must further study and perfect it. We start late on developing our own computers, but we should catch up with others."

Only when the national computer team constantly injected "new power" could it became increasingly stronger, and eventually completed the historic task of computing breakthroughs and catching up with the world.

In mid-November of 1958, Chen Geng specially listened to the special report by Ci Yungui about the development of the "901" computer and clearly indicated, "You developed the computer through learning and practicing and you need to continue to develop and cultivate people to be computer experts. A university cultivates people. It should both develop products and cultivate talents. Let the Chinese computer industry develop faster. You should establish a computing major immediately when you return. As for the source of the trainees, you can go back and write a report and go ask Deputy Commander of the Navy, Luo Shunchu, for approval."

Chen Geng made a prompt decision. The Party Committee of the Institute implemented the policy immediately and made five decisions.

First, it set up a temporary computer teaching organization in the department of Navy Engineering and appointed Hu Shouren as the responsible person.

Second, after entering the Institute, students would first participate in the practice of computer development for two months and make up for missing knowledge. They could enter the undergraduate study after passing the exam.

Third, the Institute applied three measures to solve the shortage of teachers. It transferred teachers from other majors, it selected graduates from local universities and it selected senior students in the Institute to graduate early.

Fourth, the instruments and equipment in the laboratory would be provided by the Institute and department offices, cultivating experimenters and developing experimental equipment in person.

Fifth, it would prepare a draft for teaching plans and syllabi immediately, and then continuously modify and improve them in practice.

After several months, the above work was basically completed. The first batch (more than 30) of computer professionals went to the Institute and successfully transferred to undergraduate study, which marked the official establishment of the computing major in the PLA Military Institute of Engineering.

This is the genuine source of Chinese computing education.

Although it was only trickle at the time, it continued to flow endlessly for decades and infused new blood into the Chinese computer industry. It enabled China's first transistor computer, China's first computer with integrated circuits, China's first 100 megaflops supercomputer and many other scientific miracles. A number of academics of the Chinese Academy of Sciences and Chinese Academy of Engineering and a group of scientific and technological talents were fostered for the rise of China's computer industry, which gave birth to great breakthroughs in China's computer industry and attracted the attention of the whole world.

Chapter II Cutting the Wrist Like a Warrior

The birth of the transistor prompted a technological revolution in computing.

When Ci Yungui visited London, England, he was keenly aware of this new technology and new trend. After he returned to China, he definitively suspended the ongoing research on the vacuum tube general-purpose computer project and led everyone to initiate the sprint towards the new frontier of computer technology.

They used backward domestic transistors to develop China's first "441B" transistor computer with the advanced standards of the rest of the world.

I. Sleepless Nights in London

In September 1961, Ci Yungui, the responsible person for China's first vacuum tube special-purpose numerical computer "901", the vice deputy of the department of Electronics Engineering of the PLA Military Institute of Engineering, visited the United Kingdom as a member of the delegation group. After he went to the University of Manchester, he was deeply shocked by what he saw and heard. He did not know that the speed of development of computer technology in the world was so fast.

The United Kingdom first initiated the development of the transistor computer in 1952 and took only one year to achieve its first run. In 1955, The United Kingdom debuted the second type of transistor computer. In 1956, Metropolitan Vickers in Manchester unveiled the larger-scale MV950 transistor computer.

The development of the transistor computer in America was even faster. In 1955, the United States successfully developed an all-transistor numerical computer, TRADIC, with a power consumption of 100W and a volume of three cubic feet. In 1956, the United States developed a test transistorized computer TX-O with a memory capacity of 4,096 words and a word length of 18 bits. In 1958, Philco from the United States produced a large general-purpose transistor computer, the Philco Transa S-2000. In the same year, the transistor minicomputer developed by the Automation Division of North American Aviation Inc. was also put into operation. In particular, the transistor computer 709TX, developed by IBM for the ballistic missile early warning system in June, 1960, was the strongest data processing machine in the world, and was widely used for the design of missiles, rocket engines, jet engines, supersonic aircrafts and atomic reactors.

All of these achievements made Ci Yungui aware that the "magician's hat" (the transistor) had become the protagonist in the world of computer technology. Computer technology had stepped into a brand-new era of transistors and the vacuum tube computer looked like a flash in the pan, and was fading from the historical stage.

China's computer industry started more than 10 years after

that of other countries. As for the exploration path for the transistor computer, other countries had been developing it nearly 10 years and had invented several types, while China had not even begun.

China's computer technology was far behind.

Ci Yungui felt a sense of urgency that he had never experienced before going to London.

Walking from the window, he sat down at the writing desk, opened the pen and wrote swiftly, "Dear Institute leaders..." He reported the developmental condition of computers in the world to Institute leaders and suggested that they terminate the development of the vacuum tube general-purpose computer immediately and prepare to start the transistor computer project.

When he had finished the letter, it was already two in the morning. However, Ci Yungui was still emotional and felt no drowsiness. He forced himself to lie in the bed, where he tossed and turned and found it hard to fall asleep.

Eventually, he climbed from the bed and sat down in front of the desk again. He spread the letter paper and wrote a line of regular script:

Design Scheme for China's Transistor Computer

China's first transistor general-purpose computer was kicked-off secretly at the birth place of the world's first transistor computer.

II. Innovation and Rejecting Inferior-Quality Products

During Ci Yungui's visit to the United Kingdom, he carefully observed the advanced architecture during the day and collected technical data on transistor computers through various channels. In the evening, he locked himself in the hotel to digest the materials and conceive the development plan for China's transistor general-purpose computer.

When he returned to China after two months, he not only brought a large amount of materials, he also brought back a complete transistor architecture and a design plan for fundamental logic circuit.

It was already seven in the evening when he returned to Harbin by train from Beijing. Ci Yungun had a simple dinner at home and went to the laboratory with the materials and the development plan. Standing in front of the building of the Electronic Engineering Department, he looked up the computer laboratory where there was still light. Going upstairs and opening the door, he found that all of the people from the vacuum tube general-purpose computer project team were there and had worked overtime. Some were testing, some were welding and some were charting. Everyone was concentrating on their work and were unconcerned about his arrival.

In this situation, Ci Yungui was both touched and anxious. What a good team they were. With a team like this, there was hope that China would catch up with the world; with a team like this, there was confidence that China would finish the development of a transistor general-purpose computer as early as possible. However, what made him anxious was that the vacuum tube general-purpose computer project had not stopped, but was in full play! It was flogging a dead horse and wasting national property.

Ci Yungui shook his head helplessly and said, "You have been working very hard!"

People came around Ci Yungui when they found that he had returned. Ci Yungui introduced his information and described his feelings when he was in the United Kingdom, especially the development trend of computers in developed countries such as the United Kingdom and the United States.

Ci Yungui told the people worriedly, "Overseas, countries have explored the transistor computer technology for nearly 10 years and it has become the mainstream. The vacuum tube general-purpose computer is far behind."

People felt enlightened after hearing his words, asking "How come the development of computers is so fast?"

Ci Yungui said, "We must immediately terminate the vacuum tube general-purpose computer and start the development of the new generation machine—the transistor computer."

Someone regretfully said, "The machine will be finished in one or two months and we cannot bear to stop now."

Ci Yungui replied, "We need to accept it, however hard it is, because we cannot continue to devote ourselves to outdated machines anymore. We can take what we did in the past as a lesson."

Everyone came back depressed and the laboratory became quiet suddenly. Ci Yungui stood still, staring at the vacuum tube general-purpose computer that was about to finish assembling and felt his heart bleeding. This machine, which represented the hard work of his comrades for more than a year, was about to be terminated. It was like a pregnant woman who was about to give birth to a baby after nine months but suddenly had to induce labour. As the subject host, he was like the mother going into the surgery with intense sorrow. He loved the "child" more than anyone; meanwhile, he was also clear that although it was a top-ranking machine in China and advanced compared to the "901", it was still an inferior-quality machine which was surpassed by others in the world. It was a congenitally deficient "child". Since this was so, he had to give it up, no matter how heartbroken he felt. He must terminate it with the determination and bravery of a warrior cutting his wrist, or it would be burdensome when moving forward and greater waste for the country.

The first night after Ci Yungui came back to the Institute was another sleepless night.

Someone said that domestic transistors not only had very little output, but they were of too low quality, and could not be used to develop computers at all. However, foreign countries were not allowed to export transistors to China and they could not be bought in the international market. If even the relatively mature vacuum tube computer was not pursued, the exploration of China's computer technology would be totally out of breath.

This was the truth. China's transistor industry started in the late 1950s and the products were extremely immature. In 1959, there was already one company starting to use domestic transistors to develop special-purpose computer. After hard work for two years, they found that the performance of the computer was unstable. Failures occurred every several minutes. Sometimes it was that the tube was burned out, or that the circuit was out of order.

Some experts affirmed that "China would not develop a general-purpose computer within five years!"

Ci Yungui did not agree. He thought that the transistor computer had already become the mainstream in the world, and if he continued to be indecisive and did not have the courage to face the difficulty, the gap between China's technology level and that of the world would get bigger and bigger. Eventually, it would affect the development of the whole national economy and the modern construction of the national defence and the military. As for the failure on the exploration of the transistor computer by the associate organization, it should be treated as though a child learning to walk had fallen down—it should not be concluded that he would never learn to walk. In addition, if no one uses domestic transistors, then how would they develop? Through his deep investigation and analysis, he thought that although some problems still existed in the domestic transistor, if they could strictly choose and design the circuit, the transistor computer could be developed at last.

At this critical moment, the leader from the COSTIND and the Institute supported Ci Yungui's pioneering work.

General Nie Rongzhen, who was in charge of COSTIND, heard the report by Ci Yungui, and then instructed the relevant departments to send a batch of rare transistors to the Institute at the time.

Liu Juying, dean of the Institute, said to Ci Yungui, "Do not worry about the money. The Institute supports you—just go ahead."

Ci Yungui and his team members quickly completed the development demonstration report of the transistor general-purpose computer. The Institute submitted the report to the COSTIND for the first time. The report demonstrated that China's computer technology had entered the era of the transistor.

III. Youth is Creation

Not only did COSTIND quickly approve the argumentation report of the Institute, but Marshal Nie Rongzhen, who was in charge of COSTIND, specifically instructed that "The PLA Military Institute of Engineering shall develop the general-purpose transistor computer as soon as possible!"

In order to implement Marshal Nie's requirement of "as soon as possible", Ci Yungui started to recruit personnel in a big way. On 5 March 1962, four months after Ci Yungui returned to China with the delegation from Britain, a general-purpose transistor computer design group, composed of a dozen young instructors, was set up in the computer teaching and research department of the Department of Electronic Engineering. Department deputy director Ci Yungui directly took up the leadership role, and Liu Dezhen served as the group leader. Kang Peng and Ji Qixian worked as the deputy group leaders. The model code was named “441B”.

However, the design group did not have a studio. Ci Yungui encouraged other people to look for a studio; later, they found that the Institute’s tank warehouse was empty and particularly spacious. Ci Yungui undertook numerous efforts to persuade others to let him borrow it. Everyone helped to clean this warehouse, prepare wires, connect the power supply, and build the workbench, quickly turning it into a computer lab.

Ci Yungui knew better than anyone about the difficulties that project “441B” faced. Although some comrades in the research and development team had been trained by the vacuum tube special-purpose computer “901”, they were still very unfamiliar with making transistors, just like children learning to walk. Moreover, the transistors in China were still in the trial production stage, and the quality was indeed poor. Using them to make computers is was like asking children to learn to walk in the mud.

Whether the team could design advanced basic circuits with backward transistors was not only the key to the development of computers, but also the biggest difficulty. Ci Yungui was determined to concentrate on solving this problem, and gathered a group of strong and capable personnel.

Ci Yungui said, “The basic circuit design of transistor computer depends on you, Kang Peng!”

With heavy responsibilities on his shoulders, Kang Peng was sent to the Institute of Computing Technology of CAS, the Institute of Computing Technology of the Fourth Ministry of Machinery Industry, and other professional computer research units.

How could they produce the advanced computer circuits with backward domestic transistors? Kang Peng was dedicated to solving this problem.

One day, he went to the auditorium to listen to a report with everyone else. As usual, when he sat down at the auditorium, he consciously or unconsciously focused on the circuit design, and had no idea who was sitting on the stage, who was lecturing and what he was talking about. In other words, he was immersed in the colourful space of a transistor circuit. The auditorium echoed the impassioned voice of the lecturer, but in Kang Peng’s mind he seemed to be walking in a pristine forest. In such a forest, everything is quiet, the sun rays go through the thick branches and leaves on the grass, and the surrounding paths crisscross. He looked down and explored the path that could lead to the vast world. Gradually, his eyes grew brighter and brighter.

The poor quality of domestic transistors at the initial stage was mainly due to their short service life, as they burned out easily, resulting in machine failure. The reason for the transistor's short life was that the power consumption was too high. To extend the transistor's service life, the power consumption had to be reduced. This is like a person who can carry a load of 50kg, but can go further if he just carries a load of 10kg.

The auditorium burst into thunderous applause. The meeting was over and everyone left, but Kang Peng was still quietly sitting as a military man.

A colleague knocked his shoulder and said, "Kang Peng, the meeting is over!"

As if wakening from a dream, Kang Peng said, "I found it!"

His colleague was confused. "You found what?"

Ecstatic, Kang Peng said, "I found the best design of transistors for a computer circuit design!"

Kang Peng applied the isolation blocking technology of the magnetic core transistor blocking oscillator, which effectively reduced the transistor power consumption, and then he invented the most basic unit of the digital computer: the isolation-blocking intermittent oscillator, and the isolation-blocking push-pull trigger.

When Marshal Nie Rongzhen learned that Kang Peng's isolation-blocking intermittent oscillator and isolation-blocking push-pull trigger were crucial to the development of the first transistor general-purpose computer, the "441B", he happily said, "We are just fighting for the Chinese. We don't want this invention to be named as a foreigner's name, but it should be named after Comrade Kang Peng!"

Li Zhuang, the director of a certain bureau of COSTIND, suggested, "I think we should name it as Kang Peng."

Marshal Nie said, "Great, let us call it the 'Kang Peng Circuit'!"

Nie personally issued the "Kang Peng Circuit" invention certificate, which was unprecedented in the Chinese history of science and technology.

After the main obstacle has been successfully broken through, other directions such as software design, storage system, input and output system, etc., steadily advanced, and the eight-bit arithmetic logic unit (ALU), 20-bit ALU, and magnetic core memory model machines were developed successively, whose function, performance, stability, and process were confirmed by running for a long time. In

July 1964, the "441B" was assembled.

Finally, the testing began. After hearing the news, two advisers were sent by COSTIND to observe. The power switch snapped on and the "441B" buzzed softly.

One hour, two hours, three hours, then 57 hours passed, and the machine never malfunctioned.

The two advisers of COSTIND said in surprise, "That is the highest record for the most famous machine in the world today. The '441B' is amazing!"

Eventually, 58 hours, 72 hours, and 332 hours went by, and the machine still functioned—that was more than five times the record of the world's best model operating without malfunction!

It broke the prediction that domestic transistors would not be able to be used to build the world's advanced computers in the short term.

COSTIND also made the collective first-class contribution for the development team.

On 26 April 1965, 26 experts from 10 units across the country witnessed the excellent performance of the "441B". It had the following features:

1. Word size of binary: 40 bits;

2. Single address instruction system;

3. Parallel floating-point operation;

4. Magnetic core memory with a capacity of 8192 words × 20 bits;

5. Clock frequency of 50 kHz;

6. Stable and reliable function. The power supply deviation range was ±20%, and the demands for power supply and environmental conditions were low.

The comprehensive level of the "441B" met the standards of the best domestic and international machines.

The team, led by Ci Yungui, using China's components and parts that were behind the standards of the rest of the world, spent three years to achieve the results it took the United States and the United Kingdom over 10 years to achieve!

The success of the "441B" was like a catalyst. From 1965 to 1970, China showed a trend of numerous players competing against each other in terms of the development of transistor computers.

After the "441B", the PLA Military Institute of Engineering adopted the "441B" technology to launch the "441C", "441D", "57-1" and other transistor computers;

The "123", "108b", "320", "850J", "108C", "801", "106-I-2" transistor computers were developed by the North China Institute of Computing Technology;

The “S-3”, “HY-Z” and “DY-4” transistor computers were developed by the 706 Institute;

The “109C” and “717” transistor computers were developed by the CAS Institute of Computing Technology.

The East China Institute of Computing Technology developed the “X-2” transistor computer;

The 56 Institute developed the “T100” transistor computer;

…

The overall level of Chinese computer technology successfully ranked among the world’s best.

IV. The Meritorious Statesman of the "Nuclear Bomb, Missile and Satellite" Plan

A proverb says that it is necessary to have effective tools to do good work.

Advanced computers are the effective tools of developing strategic weapons. In other words, to solve the mathematical problems in strategic weapons tests, China must first have advanced computers.

Since their establishment, several strategic weapons bases located in the depths of the desert in northwest China, and a conventional weapon testing base located in the hinterland of the northeast Horqin prairie, have been equipped with only one vacuum tube computer, which is huge in volume, complex in operation, difficult to use, and slow to calculate; it can only complete dozens of operations per second and cannot solve complex mathematical calculation problems at all. The base officers had long looked forward to a high-performance computer, just as the dry desert longed for a clear rain.

In April 1965, as soon as the "441B" passed the national appraisal, COSTIND instructed the PLA Military Institute of Engineering to replicate three machines to meet the urgent needs of these three bases. The researchers fought day and night to complete the replication task and doubled the calculation speed of the machine, achieving more than 14,000 flops. When the machine was sent to several strategic weapon test bases and missile test bases, the officers in the bases happily hid it as a treasure in the machine room; only the scientific research personnel who needed to solve the problem and the top leaders from the central government could enter the machine room. For others, there was no chance for them to even just take a look at it. The "441B" was the base's "busy man", operating for almost 24 hours a day, constantly reading the various complex mathematical problems encountered in the research and development of nuclear bomb, missile and satellite.

The "441B" system, which was equipped in the Horqin conventional weapons test base in 1966, dealt with a series of complex data in the set of compiled data for various conventional weapons firing. In October 1982, the base welcomed a group of experts from the National University of Defense Technology who had come to study and made arrangements for everyone to visit the computer room. When everyone entered the computer room, they saw the "441B" cabinet neatly arranged in the room. The set of boards looked like books arranged in an orderly manner, and the red and green lights on the machine twinkled mysteriously.

Lu Zaide, an old engineer, said during the visit, "Let me introduce you to the greatest contributor of our base."

Everyone laughed when they head such words, making Lu Zaide confused.

At that moment, the leader of the university research team pointed to an expert and said: "He is Kang Peng, the greatest contributor, and chief designer!"

Lu Zaide held Kang Peng's hands and said: "Professor Kang, I've heard so much about you. This computer you've built is simply amazing. The computers outside have been updated for generations, but we are reluctant to replace it, and we have been using it all the time."

Kang Peng asked excitedly, "What tasks do you use it to carry out?"

Lu Zaide said, "Firing table calculations, film theodolite

interpretations, rocket remote sensing processing, and photographic data processing. Almost all complex data processing depends on it."

The computers are generally updated once within five to six years, and "441B" users have generally been using it for more than a decade, or even more than 20 years!

On 28 December 1966, two years after China's first atomic bomb was exploded, China's first hydrogen bomb was successfully tested. Six months later, at 8:20 a.m. on 17 June 1967, a military plane dropped a hydrogen bomb equivalent to 3.3 million tons of TNT on Lop Nor; another loud noise came from the Gobi Desert in the northwest of China, declaring the perfect success of the China's hydrogen bomb test.

On 24 April 1970, the "Long March 1" carrier rocket rose from the Jiuquan Aerospace City and sent China's first man-made earth satellite "Dong Fang Hong I" into space, and the beautiful "The East is Red" music echoed in the vast universe.

On 26 November 1975, China successfully launched its first recoverable remote sensing satellite with the "Long March 2" carrier rocket, and successfully recovered it on 29 November. China had become the third country in the world to master the technology of satellite recycling. Since then, the "Long March 2" series of carrier rockets have successively carried a series of recoverable remote sensing satellites into space.

At 10:00 a.m. on 18 May 1980, a full-range intercontinental ballistic missile "Dongfeng 5" was launched from the Jiuquan launch site into the high seas south of the Gilbert Islands in the central Pacific Ocean. The missile flew for about 30 minutes, with a ballistic peak of more than 1,000 kilometres, and a range of more than 9,000 kilometres. China was the third country in the world to possess the intercontinental missiles after the United States and the Soviet Union.

…

Before the 50th anniversary of the founding of the People's Republic of China, the CPC Central Committee, the State Council and the Central Military Commission awarded the "Nuclear bomb, Missile and Satellite Contribution Medal" to 23 scientists, including Qian Xuesen, Zhu Guangya, Ren Xinmin, Cheng Kaijia, Wang Daheng, Yu Min and Chen Fangyun, in recognition of their outstanding contribution to the "Nuclear bomb, Missile and Satellite" project.

If there is a medal to be given to a device making outstanding contributions to the "Nuclear bomb, Missile and Satellite" project, then the "441B" should be given this medal!

Chapter III
The Great Man Assigned Officers Tasks

Deng Xiaoping, on the first day of work after of his official announcement that he would return, heard the report by the leaders of the Changsha Institute of Technology (CIT; the predecessor of the National University of Defense Technology), and felt happy knowing that the computer capable of megaflops was developed by a team under the leadership of Ci Yungui.

At the National Science Conference, Deng Xiaoping, the chief architect of the "reform and opening up" policy, pointed out with great foresight that "China wants to carry out the four modernizations, but it cannot be done without the supercomputer!"

At the argumentation conference of China's first supercomputer of 100 megaflops, Deng Xiaoping personally ordered that this project was to be given to COSTIND. The Institute of Computing Technology of CIT of COSTIND produced the computers capable of megaflops in the 1960s and 1970s, making it a highly capable team.

I. The sound of spring

Throughout the world history of the development of the computer, the process of computer replacement and upgrade development is like a competitive relay race; when the player with the first rod is sprinting, the next player is already running, to complete the handover of the rod.

Seymour Cray, known as the father of the world's supercomputers, is undoubtedly an excellent receiver of rod in the exciting relay race of computer technology. In the early 1970s, Cray began to design supercomputers using the new technology of vector quantity supercomputing, and in 1975, he launched his painstaking work, the first true supercomputer, the "Cray-1", in human history.

Although it is known as a supercomputer, its body is not large, covering less than seven square metres. It contains a total of 350,000 integrated circuits, and has a weight of no more than five tons. Its calculation speed was dazzling at that time—it could maintain 100 megaflops sustainably.

In the years when Cray devoted himself to designing supercomputers, Ci Yungui was still observing the trend of international computers closely. In 1972, he proposed the idea that China would build a supercomputer, and he travelled to different places to observe them.

After listening to Ci Yungui's words, many people were surprised, saying "Under the current conditions of China, it is even difficult to produce a computer that can maintain megaflops--is the ideal of producing a supercomputer capable of 100 megaflops just a fantasy?"

His colleagues lamented: "We can never keep pace with the ideas of Professor Ci."

Indeed, Ci Yungui's vision of exploration is always far ahead of others. At the beginning of the 1960s, China was developing vacuum tube computers at full blast, but he stopped his research projects and travelled around for the transistor computer project. In the late 1960s, the technology of domestic transistor computers had just matured, and he also proposed producing computers with integrated circuits, determined to achieve a large leap from 10 kiloflops to megaflops.

At that time, the "151" central processing unit of megaflops to be installed into the "Yuanwang 1" surveying vessel was in the tight phase of design. Someone told Ci Yungui, "We don't know if the '151' can feature the megaflops—we'd better finish what's on our hands, and then we can consider the supercomputer. If we want both, maybe we will achieve nothing."

After listening to such words, Ci Yungui, who had heard enough criticism, and learned to be silent, argued, "We not only will achieve both, but also achieve more. If you just focus on what's on your hands, then you will miss the future opportunities; after finishing what's on your hands, you will have nothing to depend on. The countries that are far ahead of us are now chasing one another, and competing against each other. If our eyesight is limited, then there will never be a place for our Chinese technology in the realm of the world's computers!"

In 1974, the report of supercomputers of 100 megaflops, combining Ci Yungui's achievements of more than one year, after being revised by COSTIND, was sent to Deng Xiaoping, who soon gave his consent to the project.

In October of this year, Zhang Aiping, Director of COSTIND, instructed the well-known computer experts in China to set up a special investigation and research group led by Ci Yungui, to travel to various cities to investigate the demand for supercomputers, domestic electronic components, and external equipment production levels.

Deng Xiaoping, on the first day of his official announcement of return to work, heard the report by the leader, Zhang Wenfeng, in Changsha Institute of Technology (formerly known as the PLA Military Institute of Engineering), on 23 July 1977.

Deng Xiaoping chatted with him for a while, then introduced the subject into the conversation, saying "You are all seniors from the PLA Military Institute of Engineering. What is the situation in Changsha now? I'd like to listen to your opinions." Zhang Wenfeng began to report to Comrade Deng Xiaoping in accordance with the prepared ideas. Deng not only listened very carefully, but also from time to time asked about the situation, made some brilliant comments about education, making people feel that he was concerned and familiar with the national and military education, and science and technology. When Zhang Wenfeng reported that the university had developed the "151" central computer of megaflops installed into the "Yuanwang 1" surveying vessel, and was planning to produce a computer of 100 megaflops. Deng was very happy and waved his hand and said, "Don't say 100 megaflops, 10 megaflops would already be good enough, but it must feature stable quality, and smaller size. You must tell the truth. If you can do it, just say it. But if you can't do it, then don't brag. For a university dedicated to scientific research, the teaching and research cannot be separated from each other, and only by doing a good job in scientific research can the quality of teaching be further improved."

In September 1977, Zhang Aiping again instructed Ci Yungui to select the technical backbones to investigate the development of supercomputers in depth, and form a report on the basis of the investigation. After summing up the two investigation reports by Ci Yungui, COSTIND submitted the report on the development of supercomputers to the CPC Central Committee on 14 November. On the 26th, the CPC Central Committee approved the report.

In March 1978, the Central Committee convened a meeting on the development and deployment of supercomputers. Deng Xiaoping attended the conference and personally ordered that "the project of 100 megaflops computer shall be given to CIT." He suggested that it was hard for CIT's Institute of Computing Technology to produce a computer capable of megaflops when the Institute was relocated to the south, proving it was a highly capable team.

Deng Xiaoping looked at Zhang Aiping and said, "The project of the 100 megaflops computer is given to COSTIND, but you have to set up a military warrant."

Zhang Aiping, Founder General, stood up, and pledged to Deng Xiaoping in a firm voice: "100 megaflops without any loss! Six years without one day of delay!"

After the meeting, Zhang Aiping called Ci Yungui to the office and said, "I have set up a military warrant for Comrade Deng Xiaoping, and you must also set up a military warrant for the Central Committee of the Party."

Ci Yungui was excited, and then said in a loud voice: "I promise: 100 megaflops without any loss! Six years without one day of delay! And nothing over-budget!"

II. An Ambitious Promise

China took its next step in the field of science and technology when Ci Yungui started developing a supercomputer with a speed of 100 megaflops.

In 1978, the CPC Central Committee and the State Council of the People's Republic of China were gathered for the important occasion of the National Science Congress. The conference had lasted for over a week from 18 to 31 March. At the opening ceremony attended by 6,000 people, Deng Xiaoping, the highest-ranking leader of China at the time, made an important speech, calling for "great ambitions and tenacious efforts for the modernization of science and technology." He made it clear that "the crux of the modernization of China is the mastery of modern science and technology" and that "intellectuals are part of the working class." In addition, Deng reiterated that "science and technology are the primary productive forces" fundamental to Marxism. In this way, Deng was able to clarify the theories or issues that had long impeded scientific and technological development, thus liberating those intellectuals.

At the National Science Congress, the CIT's Institute of Computing Technology received the "National Science Congress Award." During the discussions at the conference, Deng went so far as to mention the development of computer technology, which made Ci Yungui even more delighted. With a long-term vision and plan, Deng said, "Supercomputers are necessary for China to carry out four modernizations!"

Having a shared his opinion with the national leader who made the powerful speech, Ci Yungui could not help thinking of the shouts and yells when he led his research team not long ago.

The designer of the national strategic weapons told him that although the experiments on strategic weapons were a success, the environment was severely contaminated due to the fact that the test explosions were conducted on the ground surface. In fact, Chairman Mao had instructed that the tests of strategic weapons be transferred from the ground surface to underground. The designer continued to say that as the difficulty of processing test data was increasing, the current computers, with a speed of 100 kiloflops and those with a speed of one megaflops, could no longer do their jobs. Besides, the United States had set about creating new test models for its strategic weapons by utilizing advanced computers. Under such circumstances, this technological gap between America and China would be further widened, unless China began to develop supercomputers on its own.

Comrades from the space sector told Ci Yungui that rocket testing cost too much. Every test would cost over tens of thousands, hundreds of thousands or even hundreds of millions of yuan (the monetary unit of China). It was often necessary to launch a rocket a few times or even over 10 times to verify a model. If China could, like the United States, have supercomputers to conduct experiments with the rocket launches, the number of tests required would be greatly reduced, thus cutting the costs.

Leaders from the aviation industrial sector told Ci Yungui that whenever they designed a new aircraft, it was required that they conduct wind tunnel tests to collect aerodynamic data, which cost the country millions or even tens of millions of yuan. The leaders from the aviation sector wished that China could develop a supercomputer that would be used in flight simulation tests, which would in turn help reduce the huge investment from the country.

The seismological departments kept complaining to Ci Yungui that the world had entered a period of ever increasing numbers of earthquakes. However, earthquake forecasting was a great challenge. They had usually hit before anyone could capture any early signs or warnings. The common people blamed the seismological departments for their uselessness. If a supercomputer could process seismic data, the seismological departments could better predict earthquakes.

Comrades from the national statistical office informed Ci Yungui that China and Japan had been equal in economic growth back in the 1950s. However, by the 1980s, Japan had been included among the strongest economies in the world, while China was far behind on the list of developing countries. Japan's fast economic development was indeed a result of various factors. But one of the most important reasons was that Japan had seen the creation of transistors as an opportunity back in the 1960s. Having seized this historic opportunity, the development of the electronics industry, based on computer technology, became the central focus of the Japanese government, giving Japan's economy a great boost.

Staff from the flood control sector told Ci Yungui about their incapability to make mid- to long-term predictions for complex climates. Flood control still remained at a rather basic and passive level without any advanced technologies involved. It was impossible to make any predictions and assessments about flooding, which caused severe damage to the lives and safety of the Chinese people and to the national assets.

A director from a geological exploration institute of the oil sector pled with Ci Yungui to develop supercomputers at a faster pace. Without self-developed supercomputers, China had to send data about oil and mineral resources to the United States every year by plane in order to conduct three-dimensional data processing. Spending a great amount of money to offer a nation's top secrets to others was indeed an immeasurable loss.

With these words, the director took Ci Yungui and some others to a hall in which a specifically decorated room stood. The director said, "This machine room belongs to the research institute."

Ci Yungui thought it was strange and asked, "Why don't you go ahead and install the machines in this hall? Isn't it unnecessary to build yet another room inside this building?"

With a grimace, the director said, "It is required by foreigners, so we have to do it."

"Is it an imported machine?" asked Ci Yungui. "Who is the exporter? How fast can it run?"

The director answered, "This computer has a speed of four megaflops. It was imported from the United States three years ago."

Ci Yungui said, "The 'Cray-1', a supercomputer with a speed of 100 megaflops, came out three years ago. Why didn't we bring in this new version?"

The director made complaints incessantly. "Professor Ci Yungui, it was too difficult. We wanted to import the 'Cray-1' from the very beginning, but they declined our request. After endless efforts and negotiations, America finally decided to sell us this supercomputer with a speed of four megaflops, but only if we accepted its terms."

Ci Yungui asked, "What are the terms?"

The director counted with his fingers, "First, we have to build a particular machine room for the supercomputer; second, the staff who use the machine and the maintenance crews will be all dispatched by the American company; third, the Chinese staff who calculated and analysed a variety of data have to stay outside the machine room and submit these data to the American staff to run."

"Plainly speaking, those terms are telling you, 'You will be grateful and pay me for the fact that I'm going to steal your country's top secrets, and you will never set a foot inside the machine room, not even to take a look at the machine.' How stringent and bossy those terms are! But you have to say yes. Because it is you who need this machine anxiously, but you do not possess the technology yourself."

Ci Yungui felt a surge of heat rushing to his brain.

At the second supercomputer research seminar, Ci Yungui sent out a firm and clear message, "This year is my 60th birthday. I promise to develop our own supercomputers no matter what it takes!"

III. "The Imperial Sword"

The Institute Party Committee had established a leading group and an office directed by Su Ke, strengthening the leadership of this project. COSTIND had also asked a task force led by deputy director Zhang Zhenhuan to be stationed at the Institute. The development of supercomputers was a huge project initiated by Deng in person, who appointed Ci Yungui to take this important task. Holding this "imperial sword," Zhang Zhenhuan tried to remove every obstacle ahead of Ci Yungui.

Ci Yungui pointed out that "The primary problem is manpower. We don't have enough technicians, not to even mention trained and skilled ones."

"You can investigate a bit more. Whoever you need, I'll bring them," said Zhang Zhenhuan.

"I have heard that the Seventh Ministry of Machinery Industry has a research institute in Hunan Province. A crew of experienced computer professionals are available right now. Can I borrow them for a few years to do research here?" asked Ci Yungui.

"Why do you say 'borrow'? In that way, how will they be able to settle down in their jobs here? How about simply transferring them to your research institute?" asked Zhang Zhenhuan.

"That would be great!" Ci Yungui broke gently into a smile. "But..." It seemed that he had something to say, but remained silent at last.

Understanding the bitterness behind that "but," Zhang Zhenhuan took a firm stand and told Ci Yungui, "There will be no reports and complicated procedures! The only thing you need to do is to give me a list of names. I will then ask for the leaders' approval. We'll see which unit dares to say no. We have to do this, if we want to achieve our goal! From now on, if you want me to do anything, you don't have to either write a report or a letter. A piece of paper with just a to-do list will do. I will finish the tasks by the time you tell me to. You can call me directly, bypassing my secretary. If it is during the middle of the night, you can call me at home. I will get up and pick up your call."

Ci Yungui had an urgent name list prepared immediately. Zhang Zhenhuan went back to Beijing with this name list and got it approved at once by director Zhang Aiping and political commissar Li Yaowen. He added three "+" on the document before handing it to his secretary to carry out the specifics. The secretary went to various departments with Zhang Zhenhuan's instructions immediately when he saw the three "+". He knew that instructions with three "+" added by Zhang Zhenhuan were the most urgent matters and, therefore, any delay would be unacceptable.

After a month, Zhang Zhenhuan gave Ci Yungui a call and asked, "Have those people arrived?"

"They have not come yet, but I've heard that they will arrive soon," answered Ci Yungui.

After hanging up, Zhang Zhenhuan rushed to the departments and lost his temper. "Now we have to simplify the procedures. Procrastination is not allowed. The time for formal statements is over. We will do whatever will benefit the achievement of this task. If we don't do so, when in the world we will be able to develop a supercomputer that can run over one gigaflops?"

A month later, all 21 computer technicians, along with their families, from the Seventh Ministry of Machinery Industry in Hunan

Province arrived at National University of Defense Technology.

Ci Yungui was overjoyed by such a fast process, but Zhang Zhenhuan was still not satisfied. “It has taken us two months to transfer dozens of people. With such inefficiency, when will we see the modernization of China?”

In this way, Zhang Zhenhuan immediately brought nearly a hundred computer technicians to Ci Yungui’s research centre. Meanwhile, after receiving some training, 50 soldiers also took part in the development of supercomputers as assistants in the research centre.

Changsha, the capital of Hunan Province, had been known as one of the “three furnaces” of China. The temperature in Changsha broke the history record that summer. It was after midnight, but Zhang Zhenhuan could not fall asleep because of this hot and sultry night. Wearing his vest, he walked to campus with his hands behind his back. Before he knew it, he had already been around the computer lab. The lights were all on inside, so he entered the lab. Right before his eyes, Ci Yungui and other experts were still working, shirtless and sweating.

Zhang Zhenhuan walked towards Ci Yungui and asked, “Professor Ci Yungui, the room is like an oven and you are all sweating a great deal. Why don’t you use electric fans?”

“Once we turn on the fans, the drawings will be blown away. We can't work then." answered Ci Yungui.

Zhang Zhenhuan said, "Then you can use the air-conditioner instead. It's scorching weather! And you are still working overtime. What if you got heatstroke? Wouldn't that delay the work even more?"

Ci Yungui was a bit concerned and said, "The entire school has no air-conditioning. I'm afraid that people will judge us if we get one installed."

Zhang Zhenhuan said, "Well, your team here is the only one developing supercomputers. It's necessary. It's not for enjoyment that we buy an air-conditioner. It's for higher working efficiency. If people are going to judge, leave it be. Once we create a supercomputer, they will have nothing to say."

With 10 air-conditioners installed, the working environment was greatly improved. People worked even longer hours and those who were still single decided to live in the lab, enhancing the working efficiency considerably.

Zhang Zhenhuan often told people around, "We have to respect scientific researchers, make friends with them, care about their daily life, work and solve difficult problems for them, and willingly take the risk for them so that they will be able to do whatever needed to be done."

IV. Seeking Speed from Innovation

With the personal care of the deputy director Zhang Zhenhuan and under the strong support from COSTIND, numerous computer talents from the base of COSTIND, the Ministry of Aerospace Industry, the Ministry of Petroleum Industry, the Ministry of Nuclear Industry, Fudan University, Wuhan University, Hunan University and Xiangtan University, continuously gathered at the National University of Defense Technology.

China was finally ready to meet the challenge of a supercomputer that could maintain 100 megaflops.

At the kick-off meeting before starting this challenge, Ci Yungui set aside the speech report prepared for the relevant departments and delivered the following impassioned impromptu speech, like a battlefield commander with one hand resting on his hips and the other hand waving in air briskly.

"Comrades! I am just 60 years old this year and I have been doing research with everyone here for 20 years. In the past 20 years, we wanted to go all out for our research career. However, either this movement or that movement constrained us and pinned us with this label one day or that label another day. These movements and labels greatly bound our hands and feet, making it impossible for us to go all out for our research career. Nevertheless, things have been changed in a good way now. Our country has focused on economic construction. As Comrade Guo Moruo said, the spring of science has arrived; there will be no political movements again. No one will ever stand in our way. Now it's good time to roll up our sleeves and spare no effort to do research, and such a time has finally arrived!"

"Comrades! Comrades! The Central Party Committee has blown the horn to announce four modernizations. Our national project concerning strategic weapons is waiting for a supercomputer with 100 megaflops; the country's economic construction is rushing to use these supercomputers. I will fight to research and develop this supercomputer even if I need to fight with my old age and at the expense of my life!"

"Comrades! The development of supercomputers with 100 megaflops is a mission decided by Comrade Deng Xiaoping. Comrade Deng Xiaoping personally appointed us to accomplish this mission. Zhang Aiping, the Director of COSTIND, made a pledge to Deng Xiaoping. I, Ci Yungui, also made a pledge to director Zhang. What is a military pledge? For the military, it is a promise that must be practiced and fulfilled even in the expense of life! If the military pledge cannot be implemented, being sent to a military court and being sentenced will be the consequence!"

"Comrades! We are now comrades-in-arms facing the same challenge. As the old saying goes, 'we will sink or swim together'. If we succeed in developing the supercomputer, then we all will be honoured. However, if we cannot make it, everyone will visit the jail together with me!"

Ci Yungui's speech was astonishing, but the road to fulfilment was extremely difficult.

"100 megaflops" requires the calculation speed to be 100 times more than that of the "151" computer which they just developed. The supercomputer spanned across two generations of models, which is a huge leap.

Ci Yungui called everyone "to aspire speed from the world's advanced technology and to seek speed from innovation!"

First of all, they needed to focus on innovating the overall technology. As early as 1973, Ci Yungui led everyone to target the supercomputer technology of the United States at that time and began to explore a general plan of supercomputing in China. After several years of follow-up argumentation and demonstration, the thematic programme demonstration meeting held by COSTIND in May 1978 confirmed that China's first supercomputer would be a dual-processor and would be completed in two phases. At that time, Liu Degui discovered an essay introducing the United States' "Cray-1" supercomputer from a foreign academic journal. Through in-depth research and analysis, he found that the design ideas and implementation techniques of the "Cray-1" were very unique and represented the international advanced level at that time. Ci Yungui immediately decided to abandon the dual-processor programme that was completed over several years' effort and turned his attention to the design idea of the "Cray-1". With the combination of the design idea of the "Cray-1" and China's national conditions, the team redesigned the overall project for a supercomputer with 100 megaflops. After several months of hard work and with the approval of relevant departments at a higher level, the overall plan of the dual-array structure was finalized. This dual-vector display structure could not only obtain two calculation results for each beat, but could enable the Chinese supercomputer to be compatible with international mainstream models from the beginning. After that, pipelined architecture and combined-pipelined architecture technology were creatively adopted to improve the parallelism and practical operation speed of supercomputers. The use of multi-module dual-bus cross-access memory architecture reduced the probability of access conflicts and satisfied the need for data traffic during dual-array operations. In addition, compression and recover transfer instructions and interval address transfer instructions were allocated, which saved space and improved speed. These innovations have successfully transformed the information transmission of supercomputers from a "single highway" into a "double highway". A "dual track" was built for it, and numerous "warehouses" were equipped along the way as its "road maintenance staff" and "vehicle maintenance workers". In such way, even if the frequency of the master computer did not change, the operating speed could be doubled. This innovative result was a major breakthrough of the "Cray-1", becoming the mainstream technology of the year. Later, the three types of supercomputer that were launched by Japan in the 1980s all adopted this structure. Even the new system "Cray X-MP" developed by companies in the United States on the basis of the "Cray-1" adopted this structural model.

Secondly, the team needed to improve component design. The approximate-iterative methods for floating-point reciprocal operation used in the "Cray-1" were improved and adopted to simplify components structures and to shorten the operation time. Moreover, the design of a fault detection system improved the hardware reliability and maintainability. The team also developed a memory module with 64K capacity and a clock cycle of 400ns. Based on the characteristics of the external machines, corresponding hardware interfaces were designed. In addition, the micro-program-controlled interfaces and double-buffer-controlled interfaces were independently designed in order to fully utilize the disk transmission rate.

Thirdly, the team needed to develop advanced software systems. If computer hardware is considered to be a stage, then the computer software is the performance that will be displayed on the stage. If we compare computer hardware to the human body, then the software is the knowledge and talents of the human body. At that time, computer software in China was unexplored territory. If software could not be developed as soon as possible, then the supercomputer would be like a stage without actors, or a giant with well-developed limbs but no brain. In order to complete the first supercomputer in China and give it a brain, Chen Huowang led the few software technicians to work overtime for many years and successfully built a complete software engineering specification for the Chinese supercomputers. The achievement of structured and modular programming and the development of simulator and debugging tools greatly improved the quality standard of software products, shortened their development cycle and accelerated the progress of the project. By using distributed system and batch processing system in the operating system, the supercomputer became fully featured, easy to use and highly scalable. The Vector FORTRAN (YHFT) adopted FORTRAN- 77l international standard text and achieved vector computing expansion according to its own machine features. In addition, the use of three-level optimization techniques with a vector recognizer significantly enhanced the vector recognition capabilities. Adopting an assembly language to describe all machine instructions and fully making use of the programmers' programming skills helped to make the most of the hardware features and achieved efficient program operation. In the end, the 80 categories and 292 modules of mathematical subsystem libraries not only had larger quantity compared with the

"Cray-1", but also had strong features and high precision and speed. The introduction and development of computer peripheral system software and the design of communication interface software solved the difficulty of quickly inputting tape data into the computer.

Finally, the team needed to achieve a breakthrough about maintenance and diagnosis technology. Double-ratio check, parity check and other technologies were adopted to build a comprehensive hardware fault monitoring system in the processor. For the main memory, the team applied Hamming codes that can correct one-bit errors and detect up to two-bit errors, greatly improving reliability and reparability. The establishment of a multi-level diagnosis system provided the necessary means for the diagnosis of problems and helped to develop error-generating software at all levels. Moreover, the development of double-ratio test benches and the main computer memory test benches greatly improved the reparability.

In addition, they also innovated high-density assembly processes, as well as an efficient uniform ventilation system with parallel and short air flow ducts. They also developed a computer-aided design system, and for the first time in China they realized the application of a highly efficient and highly reliable parallel power feed system with a multi-phase rectifier filter with AC regulated voltage and DC unregulated voltage.

At that time, the quality of domestic components was poor and the level of technology was extremely backward, which brought serious challenges to the quality of supercomputers with 100 megaflops.

Ci Yungui insisted, in his words, that they "seek quality from precise management and strict control!"

In order to ensure that the supercomputers could operate stably and reliably, the quality control and management personnel had to be meticulous. For example, the bottom soleplate of the computer machine had around 25,000 winding wires and 120,000 winding points, which they inspected more than eight times. The whole machine had more than 800 multi-layer printing plates, and each plate had 5,000 metallization holes. They checked and inspected the wall for each hole and performed hole continuity tests and insulation tests. The whole machine had more than 600 plug-in plates, and each plate had nearly 4,000 soldering points. They created the miracle of having more than two million soldering points on one machine, and not one of the soldering points were faulty. They strictly controlled the computer machine quality, adhering to the zero omission rule of the quality inspection for each component, the zero tolerance rule with respect to any defects found from the inspection and the zero concessions rule concerning the difficulty of eliminating hidden risks.

While the dual-array structure improved the computing speed of the supercomputer by two times, it also doubled the difficulty of machine design at the same time. This was especially true for the "brain" and "heart" of supercomputers, as the design of the ALU, instruction control unit and memory controllers were even more difficult. The comrades who were responsible for the design of this system were "stepping on thin ice and facing the abyss" at every step of their work. After everyone worked day and night for 15 months and finished more than 5,000 spider-net-style logic diagrams, the operation commander found that some components did not meet the design requirements during the last drawing review. Although these components were just a few individual pieces, the person in charge of this system still required that the design be "reworked immediately. The system would not work if even just one component was not qualified!" So, everyone again worked overtime day and night for more than two months. In the end, all of the hidden risks and problems were fixed, ensuring that the main frequency indicators exceeded the design requirements.

The quality of the logic design for each part of the main

computer machine was directly related to whether the computation of the supercomputer could achieve 100 megaflops. On the basis of a large number of theoretical analyses, the researchers preliminarily drew up a set of design specifications and transmission rules after repeated experiments. At the beginning of 1980, all of the logic design was completed. The researchers started their engineering design process and started to innovate the original design specifications and transmission rules one by one according to the experimental results of the circuit room and based on the new test results of the model machine. There were more than one million circuit wires for the main computer machine. They were checked and calculated one by one. The team put in tremendous effort in order to complete the task, but they found that the command control and vector registers had some minor problems when they checked the system. They did not hesitate to reinvent the scheme they had designed through hard work and tremendous effort and restarted the work. The comrades responsible for the design had no complaints. If the modification did not work the first time, then they continued to modify the second time and third time until it fully met the requirements.v

Tens of thousands of signal wires were intertwined in the

computer machine like hair that had never been combed. If one of these signal wires was wound to one wrong point, it was enough to paralyze the whole machine. The researchers said, "No matter how many signal wires there are, we have to check them one by one to make sure everything is safe." A newly graduated comrade volunteered to undertake this detailed and tedious task. In the densely packed signal wires that were piled several inches high, he carefully checked each row and compared it with the design drawings. Sometimes when he was searching and checking, his eyes would suddenly become blurred, and the winding pins would move like a group of crawling ants. Even if you manage to keep your eyes open, it is hard to avoid making mistakes. This is called "fatigue-induced error" in medicine. In order to prevent fatigue-induced errors, he stood up and ran in place whenever he felt sleepy in order to inspire his energy and spirit. When his eyes felt blurry and tired, he placed a wet towel on his eyes in order to make his eyes see clearly, but his two eyes looked like two big red apricots afterwards. At the final inspection, when the plug-in board had been placed in the computer cabinet, someone advised him, "You have already checked this by three times, and there will be no problem in the end. You don't need to check it again." He replied, "If there is a problem, then it will too late to regret it." He went down on his knees and put his head into the cabinet with a 100-watt light bulb lighting on his head. He insisted on checking the final signal wire before he got up and reported to the person in charge of the system, "All signals wires have been checked three times and all the misconnections are corrected to ensure everything is safe!"

"Six years without one day of delay!" Six years was the time that the world's advanced countries took to create a new generation of their computers. However, we were far behind developed countries in all aspects, from funding support and technical reserves, to industrial foundation and development conditions. This was like a marathon competition. Others were running on a flat asphalt road, while we were trekking in the mud. It is hard to imagine how difficult it is to reach the final destination same time with others.

However, Ci Yungui was very confident about this, saying "We are used to this situation, and we have a series of skills and methods to solve this problem."

The so-called series of skills and methods was in fact the ability to continue rushing forward while other people got tired and sat at the side of the road to take breaks, and to continue running forward while others stop to drink coffee during running. As long as their hearts are still beating, they will not stop sprinting!

In their own words, "In order to catch up with the task, we extent the working hours as much as possible and shortened the rest time as much as possible. There were no holidays and Sundays for us, and we woke up early in the morning every day and didn't leave work until late at night. We used one day as two days, and we even earnestly wished that the time could go slower. At night when it was dark, we eagerly looked forward to the daybreak, and when it was at day time, we expected the day time could last longer or forever."

V. The Yinhe-1's Great Victory

When the first spring breeze blew over China in 1982, there was good news from the supercomputer development team: the hardware debugging of its mainframe has been finished. On hearing this news, the Director of COSTIND, the Deputy Secretary General of China's Central Military Commission Zhang Aiping gladly named the first supercomputer of China "Yinhe-1".

The development of the "Yinhe-1" entered into the last phase, the joint debugging of software and hardware. Researchers further improved the software level and achieved good conventions and coordination through employing more software talents, enhancing horizontal coordination, strengthening software engineering and standardization and designing simulators and debugging tools. They also repeatedly organized trial calculations, checked hidden dangers and improved the weaknesses to further raise the machine's stability and reliability.

In May 1983, the "Yinhe-1" began inner trial calculations. On knowing this, vast users countrywide arrived at the National University of Defense Technology to witness the greatness of China's first supercomputer and try to test its capacity. At that moment, people and vehicles crowded the front of the Institute of Computing Technology. School hotels were also jammed with people.

A researcher from the Geophysical Prospecting Bureau of Oil Ministry, Zhao Zhenwen, input the seismic data processing and interpretation program into the "Yinhe-1". Soon, with a burst of melodious sound, the printer spit out several clear geological section maps. Zhao held those maps and said excitedly, "This is the first geological section map processed and printed by China's own machine!"

The famous aerodynamicist and educator Luo Shijun was studying the issue of high transonic flow. Because of its complex mechanism, it had not been solved worldwide. He asked his students to use the "Yinhe-1" to simulate the issue and the result data turned out to be the same as that published by United States.

A researcher from the Second Ministry of Machinery Industry, Wang Zhenyu, had been trying several calculations on the "Yinhe-1" for months. After careful tracking and analysis, he finally concluded that "The operating system has passed the test! This OS designed by NUDT is the only system in China that passed the test!"

…

In November 1983, the Chinese government organized the final evaluation of the "Yinhe-1". Party and State leaders, including Fang Yi, Wang Shoudao and He Changgong and scientists such as Deng Jiaxian went to the scene to witness the comprehensive, cautious and strict evaluations for the "Yinhe-1" by eight technical appraisal groups.

According to the rules, the mainframe was allowed to break down once every 24 hours; however, the "Yinhe-1" did not break down once after running constantly for 289 hours over 12 days!

Calculating one parameter of a scientific computing problem normally cost 70 hours on a computer with a speed of 300 kiloflops, but on the "Yinhe-1", it took the mainframe less than 10 minutes to fetch the first parameter.

The test experts were both astonished and confused, asking "Is it not done, or is it miscalculated?”

The data were printed out and turned out to be entirely correct.

The test expert gave an admiring thumbs-up and said, “The ‘Yinhe-1’ is really excellent in calculation.”

The “Yinhe-1” finished 26 problems of this kind, each three times, with the same result. The accuracy of all the results met the requirement.

The comparison between the “Yinhe-1” and the “Cray-1” is as follows:

Clock frequency: “Yinhe-1”, 20 MHz; “Cray-1”, 80MHz.

Computing speed: “Yinhe-1”, more than 100 megaflops; “Cray-1”, more than 100 megaflops.

Maximum storage capacity: “Yinhe-1”, four million words; “Cray-1”, four million words.

Actual storage capacity: “Yinhe-1”, two million words; “Cray-1”, 520,000 words.

I/O channel: “Yinhe-1”, 12 pairs; “Cray-1”, 12 pairs.

Power consumption: “Yinhe-1”, 25 kW; “Cray-1”, 115kW.

Software: “Yinhe-1” has an operating system, assembler, FORTRAN compiler, vector recognizer, 299 modules of 80 categories of numerical subsystem of front-end engineering platform, diagnostic program, etc.; “Cray-1” has no vector identifier and it has only 82 modules of 41 types of numerical subsystems.

Mean time between failure: “Yinhe-1”, 441 hours; “Cray-1”, 152 hours.

Time on development: “Yinhe-1”, five years; “Cray-1”, five years.

The Director of the National Accreditation Board wrote on the expertise report, “The ‘Yinhe-1’ is the first supercomputer of 100 megaflops. Its system is stable and reliable; its software is quite complete; its main technologies have met and exceeded the identification outline requirements; it is domestically advanced and in certain aspects reaches international level; its successful development fills a domestic gap!”

On 1984’s National Day, China held a magnificent military parade. The “Yinhe-1”, as China’s important scientific invention, passed slowly before Tiananmen Square, surrounded by flowers and received review from Party and State leaders.

VI. Goodbye, a Punching Bag of Depression

Since the launch of the "Yinhe-1", a large number of users have been coming in a continuous stream. The aerospace division demanded the computer to handle the problem of simulated rocket launch. The aviation sector wanted to solve aerodynamics problems of designing new fighter aircrafts. The meteorological department requested a profound scientific calculation of mid-term weather forecasts.

The leaders of Jianghan Oil Field visited eagerly with complaints: "Three-dimensional seismic data interpretation is the dominate approach for oil exploration as well as the main basis for oil reserves estimation. However, the computer we currently apply to three-dimensional seismic data interpretation is too outdated. It not only consumes plenty of time, but the errors in processing results greatly affect the design of oil field and oil exploration. Please spare no effort to help us solve the conundrum."

The Yinhe Application Innovation Team was anxious about the urgent needs of users. Working overtime, they developed the scientific and efficient three-dimensional seismic data interpretation software, accomplished 100 kilometres of three-dimensional seismic data interpretation, and provided scientific basis for the rapid and sustainable development of the Jianghan Oilfield.

On account of the laborious scientific computing tasks, the average annual delivery rate of the first "Yinhe-1" reached 92%. Therefore, it was run continuously for several days almost every time it was turned on. Not until it is too tired does it get time to stop for a while.

The operation personnel sigh with emotion: "Fortunately, the 'Yinhe-1' enjoys a stable performance and extremely low failure rate; otherwise, both the machine and us would fail to endure such arduous calculating tasks."

The second "Yinhe-1" was urgently transported to a certain geophysical research institute of the Ministry of Petroleum shortly after its production. On that day, the Institute welcomed the "Yinhe-1" as though they were greeting a hero's triumphant return from the battlefield. All the cadres and workers, including their family members and children, called out and lined the streets to extend a warm welcome.

The Institute held the closing ceremony of the imported computer room on the second day after the arrival of the "Yinhe-1". When the leaders took off the sign on the doorway indicating that Chinese personnel were prohibited from entering, the scientists present responded with rapturous applause and were moved to tears, saying "We can finally apply our own supercomputer to process data in our own land. Gone are the days when we had to send money or secrets to others, humble and slavish as it used to be."

Several young scientific technicians unrestrainedly waved their hands at the imported machine: "Goodbye, punching bag; let it go, machine of depression!"

Compared with the machine imported from the United States, the "Yinhe-1" not only improved the computing speed and combination property by several orders of magnitude, but also developed the first massive seismic data interpretation system based on the Chinese national conditions. Moreover, to address the challenges of large processing data, excessive manual intervention, simultaneous operation among users, obvious operand differences in various processing modules, and batch processing, they advanced the host of the "Yinhe-1", increased two front-end machines and corresponding software and hardware interfaces, developed a whole set of seismic application software, added some utility programs and established a network system called YHNET and a long-distance workstation (constituting a huge composite distribution computer system). All of these transformations have solved a series of conundrums, including the data transmission of different characters and word lengths among heterogeneous machines, seismic software optimization and functional distribution processing, which greatly enhances the efficiency of data processing. After the system was put into use, the machine operated stably and reliably, achieving the goal of refined seismic data interpretation and making great contributions to the establishment and analysis of the national petroleum reservoir model.

The massive seismic data interpretation system passed the national authentication in 1987. The certifying committee considered this system to have achieved a domestic advanced level. Its introduction narrowed the distance between China and other countries in the field and reached the level of the early 1980s, which was a major scientific research goal in China. It provided powerful tools for oil exploration and accumulated valuable experience for developing large-scale computer application systems in China, indicating that China had taken a huge step forward on oil exploration, development and application software.

In the same year, the massive seismic data interpretation system won the first prize of the National Science and Technology Progress Award.

The Southwest Computing Center is mainly oriented towards national strategic security applications. Since the installation of the third "Yinhe-1" at the Center, it has been fully occupied. In the past 10 years, the Center has provided more than 8,000 machine hours annually for a total of 100,000 machine hours, and the data interpreting speed has increased by several orders of magnitude, quickly resolving a series of historically important scientific calculating problems for national strategic security projects, thus effectively boosting the progress of the projects.

Experts say that "With the 'Yinhe-1', the topic that used to take several months to complete will be accomplished in several seconds now, and the accuracy has been improved dramatically; problems that failed to be solved in the past now can be easily resolved. If we initially make use of machines like the 'Yinhe-1', our development of strategic weapons can be completed several years in advance of schedule."

While completely utilizing and manipulating the popularization and application of the "Yinhe-1," the Institute of Computing Technology of the NUDT also utilized the developing technology of 100 megaflops, and worked closely with the Ministry of Electronics Industry to develop a mini supercomputer called the "MSC". Experts believe that this type of computer has a high starting point for development and new technology, indicating that China has successfully mastered the independent design and engineering development technologies of international mainstream computers. Compared with similar models in the United States, the performance of the "MSC" exceeds Convex C-1 and was close to that of the Convex C-2, filling the gap in China. It is a new model between supercomputers and minicomputers, which combines the advantages of both. It contains the high speed and large capacity of supercomputers and the acceptable price of minicomputers. It has won warm praise from small- and medium-sized users and is extensively used in the field of aerospace, aviation, energy, meteorology, C3I, CAD and artificial intelligence. The benefits these advancements have contributed to society cannot be overstated.

VII. A Monument

After the successful development of the "Yinhe-1", Ci Yungui was elected to be a member of the Chinese Academy of Sciences and was transferred to CONSTIND as an advisor to the Science and Technology Committee.

Although Ci Yungui had left the frontline of scientific research, as the judge of the computer science discipline appraisal group both at the Academic Degrees Committee of the State Council and the National Natural Science Foundation of China, he still kept up with the development of computer science in the world. He was also one of the judges to give the National Defense Science and Technology Progress Award and Invention Award. Among many other titles, he was a member of the editorial board of the Science and Technology Review and Future Generation Computer Systems.

In October 1981, the international conference on the fifth-generation computer was held in Tokyo, Japan. At that conference, Japan announced to the world that it would develop the fifth-generation computing system and introduce an intelligent computer system before 1992. This intelligent computer system resembled a human's ability to reason, study, associate and explain, and it would be even more intelligent than humans when dealing with certain problems. The creation of such an intelligent computer system would exert great impacts on every aspect of people's live and would also strongly influence social progress in the future.

This plan aroused great repercussions on the international stage. Countries with strong computing technologies introduced their own plans, one after another. America, Britain and the rest of the European Economic Community announced their development plans for the intelligent computers and competed with each other to have the best research teams.

In the face of new challenges, Ci Yungui went to visit leaders and departments that were related to the development of computer technology. Ci Yungui suggested that those leaders and departments should gather computer talents to form a team as soon as possible so that this team could tackle technology problems concerning intelligent computing. Ci Yungui established the China Computer Federation's expert group to do research on intelligent computers. At the same time, he led a group of young computer scientists and PhD students to study the topic of intelligent computing, making achievements in logic programing, parallel programming model and to meet the world's advanced level.

Ci Yungui went home late that night after attending an academic conference on computer science. After the dinner, Ci Yungui's wife, Qu Shuqin, said, "I'll walk with you."

Ci Yungui said, "The International Conference on Intelligent Computing was about to convene in America, and the deadline for submitting those papers are coming. I need to make some edits to those papers and hope to send them out tomorrow."

At around 8:00 that evening, Ci Yungui, who was working at his desk, fell forward on the desk. A week later, the superstar of China's computer technology had fallen. Ci Yungui passed away at the age of 73.

Two months after Ci Yungui's death, the International Conference on Intelligent Computing was held in America, as expected. After the opening ceremony, the conference chairman stood up and said, “Mr. Ci Yungui, a Chinese computer scientist who had made prominent contributions to the development of intelligent computing in the world, passed away a few months ago. This is not only a great loss to China but also to the world. Let us stand in silent tribute to Mr. Ci Yungui.”

All the experts who attended that conference stood in silent tribute for three minutes, expressing their deep condolences to “the father of supercomputers in China.”

Countries all over the world build gravestones for the dead who had made great contributions to the development of their nation. In order to show the people’s respect and admiration, they often write an epitaph that contains a list of merits of the individual who had died. In this way, people read these words and always remember the dead.

If we built a gravestone for Professor Ci, what would we write on it?

In fact, there was no need to write anything. He himself was a monument that existed in the minds of the people, and they would always respect and admire him. His gravestone was built with the following contributions:

China’s first special-purpose numerical computer,

China’s first transistor computer,

China’s first IC computer with a speed of one megaflops

China’s first supercomputer with a speed of 100 megaflops

...

Every one of the machines Ci Yungui had created was a monumental work in the history of the development of supercomputing technologies.

Chapter IV
Breaking the Blockade

To break through the blockade of Western countries, Yinhe researchers decided to develop a supercomputer with one gigaflops, the equivalent of one billion calculations per second, with the money they would have used to buy a foreign one with a similar computing speed.

Therefore, in 1992, the Chinese "Yinhe-2" supercomputer was developed, which ranked China right after the United States and Japan as another owner of a supercomputer with a computing speed of one gigaflops.

I. Worries of the Yinhe Researches

Some American representatives from the National Center for Atmospheric Research (NCAR) came to the China Meteorological Administration (CMA) for a technological exchange soon after the "Yinhe-1" was made known to the public. After the technological events, CMA officials arranged a city tour of Guilin for the American representatives. The representatives asked to go there by train instead of by plane, in order to stop by Changsha so that they could have a day viewing the "Yinhe-1". When they saw the "Yinhe-1", the American representatives asked to test the supercomputer in person.

People did not understand why the representatives would make this request. Some people thought they just wanted to experience something new. Some guessed they intended to steal China's technology. Still others thought the Americans wanted to test the "Yinhe-1" to see whether or not the news was true. There was no need for the Americans to learn from China, since they had much more advanced technologies than China had. Thus, the first two guesses did not make sense. Only the third one was possible, yet it was unsupported by evidence.

The truth was that some Western countries suddenly announced that they would allow the export of their 100 megaflops supercomputers to China at a discount of over 50%. Such supercomputers used to be on the embargo list to China, and even when they were traded were incredibly expensive. Now the prices of the foreign supercomputers were much lower than that of the "Yinhe-1". Their intentions were obvious—to seize the Chinese supercomputer market to nip the new "Yinhe-1" in the bud. As a result, only three "Yinhe-1" supercomputers were sold, because Chinese supercomputer buyers were going through financial difficulties in the early stages of the "reform and opening-up" policy and chose the foreign supercomputers because they were much cheaper.

Meanwhile, Western countries still held onto supercomputers

with higher performance, and forbid exports to China. At that time, the CMA wanted to buy an American supercomputer with a speed of one gigaflops to forecast the weather in the middle-to-long term. The United States charged the Chinese ten times more for the same products sold to European countries. In addition, they had many unreasonable requests, like the construction of specific rooms to house the computers, and insisting that no Chinese individuals were allowed to enter the machine room.

Chinese representatives questioned them, asking "Why have you always been giving us a hard time, even when we are here with a good intention? Why is that?"

The American representatives spread their hands and said with innocence, "We cannot do anything about it. If we do not do as the system requires, we will be punished by God."

By "God", they meant the Coordinating Committee for Multilateral Export Controls (CoCom). This organization, headquartered in Paris, was initiated in 1948 by the United States and was officially established in 1949 by the United States, the United Kingdom, France, Federal Germany, Italy, Denmark, Norway, Netherland, Belgium, Luxemburg, Portugal, Spain, Canada, Greece, Turkey, Japan and Australia, among others. Its objective was to carry out embargo policies towards socialist countries. Military weapons and equipment, top-notch technological products and strategically essential products were all prohibited as exports.

The very next year after the CoCom was set up, China was placed on its list of embargoed countries. There were 500 more goods embargoed as exports China than to the leading socialist countries, including the Soviet Union and some Eastern European countries. Among these goods, the supercomputer, of vital importance to national security and social development, was one of the most strictly embargoed.

Thus, the CMA's negotiation with the United States concerning the purchase of supercomputers went on for years but still led nowhere.

This is a consistent strategy of Western countries. They engage in "strangling" and "elbowing" to contain the high-tech growth of developing countries, in the sense that they strangle developing countries by refusing to sell equipment to them, but when they finally produce their own high-tech products, Western companies will elbow them out of the market by dumping lower-priced products into the market.

To avoid being strangled or elbowed out of the market, there is but one way to go: strive to catch up!

The Yinhe researchers knew that Chinese buyers who chose

to buy foreign supercomputers should not be blamed for the failed debut of the "Yinhe-1", because tight budgets mean purchasing lower-priced goods.

The Yinhe researchers worried about the foreseeable domino effect—the Chinese supercomputer technology would be discouraged, the gap between China and Western countries would increase, foreign products become even more expensive, the attached terms would become more oppressive, the plot to steal China's secrets would be more obvious, the "strangling" would be tighter, and in the end, the Chinese national defence would have no choice but to submit to humiliation.

This is why the Yinhe researchers made up their minds to break free from the stranglehold to develop the "Yinhe-2" supercomputer, at all costs!

At the National University of Defense Technology (NUDT), Chen Fujie, Director of the Institute of Computing Technology, Zhou Xingming, the chief engineer and Chen Lijie, the deputy chief engineer, wrote up a joint proposal to the CPC Central Committee and the State Council, saying that "in this era of high technology, some developed countries are really making supercomputers with better and better performances. If we do not seize this opportunity and strive for higher targets, we will soon be put on a back foot." By this, they proposed to initiate a plan to develop the next supercomputer with better computing abilities. COSTIND also handed in a proposal called "About Developing the Yinhe Supercomputers" to the State Council.

The CPC Central Committee and the State Council replied that "China must have independent supercomputer technologies! The State Planning Committee (the former NDRC) and the Electronic Rejuvenation Leading Group Office under the State Council would fully support related projects."

Therefore, gaining the advantage for China in supercomputing became a task for the second generation of Yinhe researchers. Among them were Chen Fujie and Zhou Xingming.

Chen Fujie was experienced in computer science. He was among the major members of the very first computer development group in the PLA Military Institute of Engineering back in 1958. He participated in several essential research projects, including China's first "901" electronic computer, the "441B" transistor computer, the "151" supercomputer that operates at a speed of one megaflops, and the "Yinhe-1" supercomputer that operates at 100 megaflops. He was aggressive and did not evade responsibilities. During the project of the "151" supercomputer, he found that pluggable storage was quite popular in the international advanced storage technology. Thus, he suggested to the chief designer that "we should apply the most advanced technology that the United States uses." Chen also pointed out how stackable storage has fallen behind the times. He said a small problem in the stackable storage would lead to more than 10 hours of downtime to open the whole equipment up, cut nearly 3,000 wires and re-weld over 10,000 spots. To avoid the drawbacks of stackable storage, China was bound to introduce advanced foreign technologies. However, someone noted, "We do not have strong technology research abilities, and the carrier rocket is to launch anytime soon. Who will be responsible if we failed to develop a pluggable storage?" Chen said immediately, "I will." Chen introduced foreign technologies and soon developed China's own larger pluggable storage to support the "151" supercomputer, which was successfully developed before long.

Zhou Xingming was a technological master who learned his ropes in long-term scientific research. He supported the development of China's first transistor computer when he was just a senior student in the PLA Military Institute of Engineering. He was diligent and innovative. He contributed many creative ideas in circuit design, logic design and commissioning. In 1965, he was deployed to the project of China's earliest integrated circuit computer—the "030" computer—to be in charge of the overall planning and integrated circuit planning. He did in-depth research in the navy and finished the overall theoretical design for the projects. At the Institute of Semiconductors in the Chinese Academy of Sciences, Zhou worked with his team and developed the diode transistor logic with strong interference immunity. In 1970, he wrote a letter to research director Ci Yungui to apply to participate in the "151" one megaflops supercomputer project, once he heard that the Institute of Computing Technology, NUDT, was taking that project. Ci Yungui approved his application, installed him as the leader of the operation control system project and gave him a difficult task—to improve the clock speed. With extensive effort, Zhou created an innovative proposal to significantly improve the clock speed to the target of one megaflops. His computer circuit was later used as the benchmark. In 1978, during the "Yinhe-1" supercomputer project, Ci Yunhui put Zhou Xingming in charge of the key technology research for the 20 MHz supercomputer. Zhou and his team developed the main body for the "Yinhe-1", which in the 1983 national technical appraisement ran for 289 hours without any malfunctions. This meant that the mainframe had reached the international standard. In 1984, Zhou succeeded Hu Shouren as the chief engineer of the Yinhe simulating computer systems. He and his team took only three years to develop China's first top-notch simulating computer system, which operated at 100 megaflops. This project won them the National Science and Technology Progress Award.

In this context, COSTIND made an exception and appropriated several million yuan in research funds to the Institute of Computing Technology, NUDT, to finance the Yinhe team on project assessment, technological preparation, circuit design, computer-aided design (CAD) development and technique preparation to get ready for the "Yinhe-2".

China has always suffered from floods and excess water. Each year, the Chinese people suffered a great loss of property or lives because there was no system to accurately weather forecast in time. Thus, in the early 1980s, CMA decided to extend its weather forecasting from a timeline of one to two days to a forecast of five to seven days.

To forecast global weather with computer technologies, it is necessary to solve a set of circulation model equations in the spherical coordinate system. Vertical height, latitude, longitude and time are needed to form a four-dimensional model. If the grid points are set 270 miles apart, a 24-hour forecast will require 100 billion operations. This means it will take 1,000 seconds even for a supercomputer operating at 100 megaflops. Therefore, if China wanted accurate weather forecasts several days in advance, China needed supercomputers with a computing speed of one gigaflops.

However, China did not have such aa supercomputer. Thus, Chinese meteorological authorities decided to import one from the United States. This is why the negotiations became awkward and fruitless.

Therefore, the CPC Central Committee and the State Council ordered the development of China's own supercomputer of one gigaflops, the "Yinhe-2", to support its medium-range numerical weather forecast.

In June 1986, COSTIND issued to NUDT a document titled "A Notice on Carrying out the Instructions of the Central Authority to Develop Supercomputers". The Institute of Computing Technology, NUDT, introduced to the meteorological centre under the CMA the initial plan of the "Yinhe-2" supercomputer, solicited opinions from potential users like the CNPC, and signed a Letter of Intent with the meteorological centre under the CMA in February 1986, before reporting to the State Planning Committee for examination.

In the Letter of Intent, NUDT said that it would provide a "Yinhe-2" supercomputer to the CMA for the medium-range numerical weather forecast system. The scalar computing speed would be at least 10 times faster than that of the American "M-170" supercomputer, the vector processing speed would be at least 1.2 times faster than the American "Cray-1A", and the general computing speed tested by the meteorological application T63L15 would match that of the "Cray-1A", which was used in the medium-range numerical weather forecast system in Europe in the early 1980s. The system availability ratio should be over 95%.

Based on the Letter of Intent and the report from the meteorological centre under the CMA, the State Planning Committee approved the expansion project of the CMA meteorological centre and listed it among the key projects of the Seventh Five-Year Plan.

In July 1987, COSTIND gathered major Chinese supercomputer users and experts in Beijing to discuss and assess the "Yinhe-2" project. They concluded that the "Yinhe-2" should be user-oriented and realize the following technological targets.

First, the "Yinhe-2" supercomputer should serve the needs of massive scientific calculation, engineering calculation and data processing in meteorology, oil, nuclear energy, aviation and aerospace science.

Meanwhile, the "Yinhe-2" supercomputer should have strong software to provide a multiprocessor operating system with interaction ability and internet access. The "Yinhe-2" should develop vector data compilation and multi-task software to support the explicit concurrent function and then solve the implicit multi-task compilation. By the time the "Yinhe-2" was debuted, a high-speed network of 4*50 Mb/s would be built.

In addition, the "Yinhe-2" supercomputer should enhance the performance of its input/output (I/O) and establish an I/O disk subsystem that operates 10 times faster than that of the "Yinhe-1". The project would need to set up an online tape subsystem which did not exist on the "Yinhe-1" to meet the requirements of the massive data I/O necessary in the interpretation of seismic data.

Finally, researchers should work closely with users to support them in building application systems and developing application software.

All of these requirements ensured that the "Yinhe-2" supercomputer would be useful to its users. Some people therefore referred to this meeting and its main points figuratively as "the channel between supercomputers and the market".

II. Impressive Achievements

With adequate manpower, Yinhe began its path towards developing the most advanced supercomputing technologies.

The head architect, Zhou Xingming, took a team to go abroad on a tour of a computer company. Zhou and his team were received with great politeness by the company's managers out of diplomatic courtesy. They brought Zhou's team to visit the beautiful plant one moment, and then took them to the advanced production workshop the next. They even went so far as to explain the principle of computer engineering to Zhou's team. However, these managers would either change the subject on purpose or remain silent when Zhou asked to visit some key places or asked some crucial technical questions.

Zhou took great interest in a small needle valve when he was observing the production line, and asked the company if he could get a sample of it as a souvenir.

To his great surprise, Zhou's request was rebuffed by the firm answer "No."

This kind of needle valve was not actually the type of secret that could not be shared. The United States government had already gone through a rigorous screening process to decide what would be presented on this tour. The company did not even make reference to the location of things that were actually worth keeping secret.

The moment Zhou returned, he determined to develop the best and fastest supercomputer, the "Yinhe-2". China would then stand at the same starting line as the developed nations at the start of the 21st century, and would impress those developed countries with its great achievements.

After the return of Zhou and his team, they immediately abandoned the plan of a uniprocessor system required by the development contract, while setting about designing a dual processor that was two times better in performance. The design then was presented immediately.

At that time, some countries had already developed a quadruple processor system with even greater performance.

Zhou's team thus also abandoned the plan for a dual processor system without any hesitation. Instead, they decided to develop a supercomputer with a quadruple processor system and proposed a general scheme that represented the advanced levels in the world.

The Yinhe team's plan of having a leapfrog development of supercomputing technologies in China was greatly supported by COSTIND. COSTIND approved the new general scheme in June 1988, claiming that the "Yinhe-2" was a general-purpose supercomputer designed for large-scale scientific research, large-scale data processing and engineering computing. With 64-bit floating-point arithmetic and a four-CPU system (applicable to a one-CPU, two-CPU and three-CPU system as well), the speed of this supercomputer could reach one gigaflops.

It had taken America a decade to develop from a vector processing to a multiprocessing parallel computer. This meant that Yinhe researchers only had four years to accomplish what America had done in 10 years. They were confident they could do it.

First, the Yinhe project had attracted a great number of computer talents, which made Yinhe an unprecedentedly strong team. The new members of Yinhe, represented by Yang Xuejun, Liao Xiangke and Song Junqiang, had become a powerful force after years of scientific research.

In addition, the successful development of the "Yinhe-1" had laid the groundwork technically for developing the "Yinhe-2". As conditions for scientific research and production had been significantly improved, there were already practical preparations made for producing the "Yinhe-2".

After the birth of the "Yinhe-1", Yinhe researchers developed a long-term vision and therefore conducted significant research beforehand on hardware and software production in order to develop the next generation of supercomputers. For example, the team had done logic simulations and delay measurements of plugins and made some adjustments in the system software, application software and tool software. All of this research had made a shortened development cycle possible. Particularly, the successful development of the YC-2000 CAD system for large-scale integrated electronics had paved the way for the birth of the "Yinhe-2".

As early as the 1980s, Yinhe researchers had already planned to develop a CAD system. The teaching and research section of the computer-aided design and measurement programme at the Institute of Computing Technology made years of tenacious efforts before they eventually developed China's own CAD system for large-scale integrated electronics. Its program volume was about 240,000 lines of statements in a high-level programming language. With adequate functions, advanced technologies, reliable performance, general applications and strong practicality, this system could be installed in any computer that was part of the VAX family or VAX stations. In this way, this CAD system could enhance large-scale computers with high speed and a high-density assembly and could facilitate the entire process of the logical design and physical design of electronics. At the same time, the engineering design of generic numerical electronics could also be improved. This achievement received the National Defense Science and Technology Progress Award.

The birth of the YC-2000 CAD system had markedly enhanced the research environment. The time of using pencils, papers and drawing boards was over.

Frankly speaking, although the research conditions were improved, there were still a great many difficulties for the Yinhe team to overcome, and there was still a long way to go.

Procuring funds was the first problem. They had only 10 or 20 million yuan, and another several million dollars USD provided by the National Meteorological Center and COSTIND for the entire development cycle of the "Yinhe-2", from pilot studies and official research to production and even the one-year after-sales warranty. The funds given to the development of the "Yinhe-1" greater than the amount offered to the "Yinhe-2", while the performance of the "Yinhe-2" was one order magnitude higher than that of the "Yinhe-1". In the meantime, the exchange rate at that time was doubled and the price index was increased by about 35% compared to that in the period in which the "Yinhe-1" was developed. In other words, the "Yinhe-2" development project was given much less in funding than was given to the "Yinhe-1" project. The price of a computer with a speed of 100 megaflops exported from an American computer company to Western European countries even represented more money than the funds given to the research on the "Yinhe-2". Yinhe thus needed to use the amount of money with which others could not even afford to buy a supercomputer to develop its own supercomputer with complete intellectual property rights.

Thus, members of Yinhe had to use these inadequate funds very carefully. They kept an eye on the market and waited until the price went down before they purchased components and equipment. They usually insisted on shopping around and waiting for the best deal. If they could produce the components or equipment on their own, they would not purchase them. As a result, they successfully developed a ground continuity tester, a plugin high-efficiency logic tester for the CPU, a main memory chip, the plugin test board and the simulation and high-efficiency test system for the disk. In addition, the development of a tape debugging system and the positioning system for a high-precision printed board had also been achieved. Moreover, members of the team had built simulators and a variety of debugging tools for the "Yinhe-2" on the "Yinhe-1", aiming to design different database software, assemblers, an operating system, a FORTRAN complier, meteorological application software, a multitasking database, test questions and other things.

With great effort, Yinhe researchers made the development of the "Yinhe-2" much more cost-effective. For example, it took only tens of millions RMB to build the four-CPU main body of the "Yinhe-2", while America spent up to 16 million dollars USD to make the main part of the "CrayX-MP/416," which was several times more expensive with only half of the capacity of "Yinhe-2".

The development of the "Yinhe-2" was a great scientific and technological challenge. It was necessary to conduct hardware, software and applied research all together. There were only about 500 or 600 researchers at the National University of Defense Technology. In addition, the CMA, the China National Petroleum Corporation, the Southwest Computing Center, the ninth and 12th institutes of the Chinese Academy Of Engineering Physics, the China Earthquake Administration, Fudan University, Wuhan University, Huazhong University of Science and Technology, the University of Science and Technology of China, the 21st base of COSTIND, the China Institute of Atomic Energy, the Shanghai Space Agency, the Second Artillery Force of the PLA and others also dispatched people to participate in the "Yinhe-2" project. The logistics departments of CONSTIND, NUDT, the Xiaofeng Corporation and the administrations of Hunan Province and the city of Changsha all took part in supporting materials supply and improving the working environment.

The administration and technological research of this project had helped each other for the making of the "Yinhe-2". With challenging and time-consuming tasks ahead, the Yinhe team focused on the main controversial issues without ignoring the less controversial ones to push forward this project with great strength.

The primary concern was collaborative planning. The project was divided into several stages, and each stage had its own timetable. The Yinhe team had also created different coordinating systems and conducted coordination efforts in almost every aspect, such as hardware, software and production. There were specific plans and instructions necessary to meet the goals of each stage. These measures guaranteed the steady and coordinated progress of this project.

The main system design and the logic design were the next important task on the list. A breakthrough in the main system design and the logic design would help software development, the CAD engineering and the making of peripheral equipment and production lines, which would drive the entire project forward.

Software verification was the next difficult problem to tackle. The verification of every piece of software was a complex and time-consuming process. Yinhe researchers built many items on VAX and the "Yinhe-1", including the assembler, different software simulators, parallel simulation environment for multitasking, parallel software development tools, program debugging tools and other necessary items. In this way, the software had been verified to be effective before it was needed for debugging on the "Yinhe-2", which greatly reduced the development period.

The tight organizational and scientific management system guaranteed the successful development of the "Yinhe-2".

When the “Yinhe-2” was being developed, finding high quality components, which had been an issue for the computer industry in China for a long time, was still a problem. As Western powers would not share high quality and advanced components with the rest of the world, Yinhe researchers could only buy components on the international market which were 2.5 times faster than those of the “Yinhe-1”. Using these components, the team was required to design the “Yinhe-2” to have an operating speed 10 times faster than that of the “Yinhe-1”.

However, members of the Yinhe team said, “No matter how difficult it is, we are not afraid. Once we start working, we will make the best supercomputer in the world.” They tried everything they could think of, such as architecture, assembly processes, software technology and design techniques to remedy the deficiencies caused by the components.

In order to achieve one gigaflops, the computer clock speed had to reach 50 MHz. That meant four CPUs had to be assembled in a space that was almost as big as the “Yinhe-1”. Therefore, they made a courageous move on high-density assemblies. Every plugin on the “Yinhe-1” was installed with 110 integrated circuits, while on the “Yinhe-2” the number was 330! The layers of printed circuit board were increased from seven to 14 layers, with its aperture size decreased from 2.54 mm to 1.27 mm. One-sided DIP was replaced with double-sided SMD in terms of components, and the most advanced vapour phase soldering was also adopted.

There were still many difficulties in the production. All the computer factories built during the time of the development of the "Yinhe-1" were clearly outdated in terms of the equipment inside. Meanwhile, the Yinhe team could not afford advanced design and production equipment. Under such circumstances, it is easy to imagine the difficulty of producing a supercomputer that was 10 times faster than the "Yinhe-1". How then did the team deal with these problems?

The answer was quite simple: the researchers solved these problems themselves. Researchers and the staff of factories worked closely together to produce models while improving their technologies. When problems occurred in the production, the technical leadership, technicians and production workers would immediately convene a meeting to analyse the cause and solve the problem. Sometimes, it would take weeks to solve a single problem. For this project, members of Yinhe had gone through sleepless and anxious nights. It took the team three years to produce the 500 printed boards which were necessary for the development of the "Yinhe-2".

There were also issues of software design. The four-CPU parallel vector processor was a new overall architecture. Thus, breakthroughs in software design were also necessary. The first problem was the new operating system. Although the operating system of the "Yinhe-1" was a good reference, the structure of the new system was quite different. The researchers had introduced many new concepts and technologies, such as multi-CPU, dual-IOU and TSS (tape subsystem), in order to reconstruct the entire operating system. Every original code had to be rewritten and compiled once again. It was also necessary for them to update the assembler, FORTRAN compiler, a variety of tools software and a parallel database. After three years of hard work, all of the programmers, who worked overtime day and night, had finished 1.09 million lines of code.

The development of the application software, which would serve as a bridge between the supercomputers and users, was yet another huge challenge in developing the "Yinhe-2". The team had given this huge task to Li Xiaomei.

Despite the fact that Li had a small figure and spoke in a soft voice with a slow pace, she was just as tenacious and aggressive as the rest of the team. In 1981, when software talents were badly needed for the development of the "Yinhe-1", Li resolutely gave away her current research project, which was almost finished, and decided to join Yinhe team. She was given the important task of designing testing questions for the two-dimensional tensor model calculation for the computer. It only took her a year to finish all the designs. After she was appointed the chief designer in 1988, she was determined to introduce parallel algorithms, a new branch of mathematics, to the field of supercomputing technology in order to further accelerate its development. With painstaking effort, Li led the team to successfully transform from vector algorithms to parallel algorithms, establishing a software system that had more than 600,000 lines of code. In this way, the throughput per unit time of the supercomputer would be markedly increased, and the parallel speedup ratio was increased by 10 times. Such improvements not only solved the application problems of the "Yinhe-2", but also effectively facilitated the nationwide application of parallel algorithms. From late 1989 to the summer of 1990, Li and the entire team went to research institutes and universities to deepen their knowledge of the practical applications of supercomputers. They collected over 150 benchmarks both at home and abroad, covering almost every aspect of the high-tech field, such as oil, energy, computational aerodynamics and computational fluid dynamics.

The system of medium-range numerical weather prediction was the primary application of the "Yinhe-2". Song Junqiang was appointed the leader for this task.

There were two challenges in front of Song and his team: first, the "Yinhe-1" was a vector processor, while the "Yinhe-2" was a parallel processor. Thus, a change from vector processing to parallel processing had to be made in terms of the programmed algorithms. Second, meteorological science was a completely new area, so they had to become meteorological experts themselves.

Song led the whole team to read specialized books about meteorology and tirelessly consulted with experts in the meteorological administration. At the same time, they had carried out comprehensive study on parallel algorithms, aiming to tackle the specialized problems of numerical algebra, numerical methods for differential equations and the computing of eigenvalues and eigenvectors. On this basis, Song repetitively studied technical documentation of over a million words that was introduced from abroad. His team carried out parallel processing and successfully made Yinhe's high-efficiency software system, a medium-range high-resolution forecast model that represented the advanced level both at home and abroad in the late 1980s. Leaders at the National Meteorological Center said that the country would save over three million RMB every year with this achievement alone.

In December 1989, the hardware design for the "Yinhe-2" had been finished, and a mini-system with two CPUs, a small main memory and a single I/O unit was put into production.

The "Yinhe-2" project had entered the last stage!

III. Leap into the Global Top Three

While the campaign of the "Yinhe-2" was in full swing, the first Persian Gulf War broke out. The information revolution of military technology, first making its appearance in the 1940s under the influence of the invention of the world's first electronic computer, suddenly emerged in this war after having lurked for nearly half a century. The NATO forces headed by the United States military, swiftly taking advantage of their advanced information-technological high-tech weapons, defeated the Iraqi army at one stroke. Their military tactics were completely new, and the might of the war shocked the whole world.

As a result of the breath-taking Persian Gulf War, people

working on the "Yinhe-2" felt a sense of urgency to build up a modern Chinese army and were quite aware of the heavy burden on their shoulders.

They thus redoubled their efforts and were determined to strive until the last moment of the "Yinhe-2" campaign, which was the moment of computer debugging.

The channel command system worked as a dispatcher that could ensure a fast and orderly transmission of various data into and out of the computer. It had taken three whole years for Fang Weirong and his colleagues to complete the design for it. But after the "Yinhe-2" had been installed with the designed channel command system for a trial, a fault signal flashed on the screen. Everybody was dumbfounded. To identify the source of the fault was like fishing a needle out of the ocean. However, Fang Weirong said: "No matter how huge the ocean is and how deep it is, I will find this needle out." For half a year, he had been working in the machine room day after day and had even caught rheumatism as a result. In the end, the "needle" was fished out and a smooth transmission between the mainframe of the "Yinhe-2" and the high-speed network was thus guaranteed.

The printed circuit board was densely covered with lines that were as fine as gossamer. There was only a distance of 0.2 mm between two lines. The whole system would collapse if any breaks, circuit breaking or adhering happened to any of the lines. Several young female workers, with the help of a projection tester modified from a conventional optical magnifier, checked every printed circuit board, and every line and spot on it, with the naked eye. The checks occurred over a summer. The heat coming from the projector bulb of hundreds of watts and from the machine made their cheeks blister, while the air-conditioner behind them sent cold air along their spines. To avoid any delay in the program, however, they worked overtime almost every night and on every holiday to ensure zero faults in any of the 20,000 pieces of printed circuit board. The task came to an end as scheduled at last, but each of the workers suffered from a sharp decline in their eyesight and had to put on glasses.

With a high sense of political responsibility and an attitude of longing for perfection, the Yinhe team quickly and steadily pushed forward their tasks. In February 1991, the debugging for the small system hardware was finished. In April 1991, the debugging for software commenced. In May 1992, the small system was successfully expanded to a hardware system with four CPUs, a main memory of 32 million words x 60 bits, two I/O processors, and a working frequency of 50 MHz. Then came the debugging for the large system. In August 1992, the debugging for the software and hardware in the whole system was finished.

Finally accomplishing in four years a rapid march covering a distance that would have been fulfilled in 10 years in normal conditions, the Yinhe team had successfully achieved a series of technological breakthroughs in the history of Chinese supercomputing.

The "Yinhe-2" was the first system in China in which four high-performance multiprocessors could share the main memory coupling system. Its basic word size was 64 bits, with a main frequency of 50 MHz and an operation speed of more than one gigaflops.

The "Yinhe-2" was successfully equipped with an independent dual-I/O subsystem that could operate at a peak speed of 30 MIPS, a memory capacity of 32 MB with bandwidth of 800 MB/S and a maximum configuration of 128 disk drives and 128 tape drives.

The Yinhe team had also successfully developed the 100 Mbps high-speed fibre-optic network and the medium-speed Ethernet interface that were in line with the international FDDI standards, which were of great help in realizing the network computing of the machine.

The high-speed multiprocessor management scheduling had been achieved in the "Yinhe-2"'s operating system, which, mainly working on the basis of batch processing, was capable of interactive processing to some degree and could support the parallel-processed high-performance FORTRAN compiler and assembly system, the multi-task database, the abundant general mathematics database, the GKS graphics library software and software tools, thus providing an effective and supportive environment for large-scale scientific and engineering parallel computing.

Based on the mainstream, general-purpose supercomputer technology, the "Yinhe-2" had manifested a series of technological innovations. These included parallel architecture, high-performance central processor design, shared high-speed main memory, an independent I/O system, dynamic fault-tolerant reconstruction for the system, three-level diagnostics, 50 MHz frequency technology, integrated CAD for high-speed large-scale electronic system, 14-layer buried-hole large-surface multi-layer printed boards, FDDI high-speed network and OSI network software, a parallel operating system and parallel and vectorized compiling. It had achieved a new level in terms of hardware, software and engineering implementation technology, all of which were at a leading position in China and some of which were as good as international advanced technologies.

The medium-range numerical weather prediction system invented by the Yinhe team had performed with high efficiency on the "Yinhe-2" platform. It only took 413 seconds to forecast the weather for each day, while the contractual time was 1,100 seconds. Having heard about the data processing speed, experts from the European Centre for Medium-Range Weather Forecasts exclaimed, "Wonderful! Wonderful!"

Software applied in the fields of petroleum, nuclear energy and aerospace and seismic research also operated with a high efficiency on the "Yinhe-2".

The trial calculation had shown that the prospect for the application of the "Yinhe-2" was quite promising. Equipped with a significant networking capability, it could be used both as a supercomputing central host and as a central processor for large-scale data processing.

On 18 November 1983, COSTIND organized experts to carry out a technological evaluation for the "Yinhe-2". The accreditation committee had made the following conclusions:

The "Yinhe-2" was China's first self-developed general-purpose gigaflops-capable parallel processing computer applicable for large-scale scientific and engineering calculations and large-scale data processing. With a stable and reliable system, its various technical indicators met or exceeded the requirements in the task statement. Its comprehensive processing capability was 10 times higher than that of the "Yinhe-1", and was equal to the international advanced level in the mid- to late 1980s. It was another major achievement from China's scientific and technological front.

The successful development of the "Yinhe-2" had narrowed the gap in technology between China and countries with advanced technologies, had once again broken the strict blockade by foreign countries in supercomputing technologies and was an important manifestation of China's overall national strength.

The invention of the "Yinhe-2" had made China the third country after the United States and Japan to master the technologies to develop a gigaflops-capable supercomputer.

Having heard the good news, General Secretary Jiang Zemin wrote an inscription for the Institute of Computing Technology of NUDT: "Conquer supercomputer technologies and win glory for the Chinese nation."

The State Council and China's Central Military Commission had sent warm messages of congratulation to the research and development crew: "Under the new situation of 'reform and opening-up', all of you have continued and carried forward the successful experience acquired as a result of developing the 'nuclear bomb, missile and satellite' program. With the spirit of self-reliance, working hard in spite of difficulties, pursuing science through pragmatism and cooperating earnestly, you have successfully developed the gigaflops supercomputer 'Yinhe-2' just within five years, which has pushed China's computer industry to a new level. It is hoped that all of you, still keeping a modest and prudent manner and continuing to make persistent efforts, earnestly conform to the spirit of the 14th National Congress of the CPC, adhere to the policy of 'reform and opening-up', promote the close integration of education with economy, science and technology, and make new contributions to the development of China's supercomputer and high-tech industries."

The successful application of the "Yinhe-2" in the National Meteorological Center has made China one of a few countries in the world capable of carrying out medium-range numerical weather prediction five to seven days out. When broadcasting weather forecasts during the primetime broadcast every day, the channel CCTV-1 uses the "Yinhe-2" as the program logo, which has been continuously displayed in recent two years. This is evidence that the Chinese people hold a deep love for this high-tech favourite and feel proud of it.

China once planned to import a gigaflops supercomputer, but the computer corporations in America not only arrogantly imposed a sky-high price and harsh conditions on the CMA, but also delayed delivery over and over. On hearing the news about the sudden advent of the "Yinhe-2", this company immediately sent representatives to China to hold negotiations with the CMA. At the negotiating table, the American delegation did not mention further conditions and even dropped the price again and again, with the only desire of signing the purchase contract as soon as possible. As a result, the purchase issue left hanging in the air for several years was solved just in a couple of days.

After the delivery by the American company, the CMA installed the imported machine and the "Yinhe-2" in the hall of the same computer room. With just a transparent glass wall between them as a partition, it seemed that they would compete with each other.

It turned out that the "Yinhe-2" had been operating stably in the competition while the imported machine suffered from faults in succession. Therefore, the leaders from the CMA had to meet with the American party to inform them of such problems. They seriously pointed out: “There is a distinct gap in quality between your supercomputer and China’s ‘Yinhe-2’. The ‘Yinhe-2’ only requires a weekly routine check, with just a few faults to be found which can be fixed very soon; but there are many troubles with your machine, and the fixing usually lasts several days. We hope that you can add more functions to your machine as soon as possible, and improve its stability and reliability.”

Hearing this news, the Yinhe team felt very gratified: “Looking into the past, we think we did not waste our life as we at least have made some contributions to the nation, though every one of us have indeed lost something.”

The “Yinhe-2” had made a great number of smart calculations and operations, as it was not only stable, but also quick to respond.

In June 1994, the Yangtze River Valley suffered from a severe flooding disaster, and three important towns in Wuhan, namely Wuchang, Hankou and Hanyang, were besieged by the devilish flood. Had the rainstorm continued, the only way to ensure the security of these important industrial towns in Wuhan would have been blow up the embankment upstream of Wuhan to channel the flood. Hundreds of thousands of people would have to be evacuated, and a great amount of farmland and a number of houses would be flooded catastrophically, which would at least cause a loss of several billion yuan. At this urgent moment, according to the medium-range numerical weather prediction provided by the "Yinhe-2", the National Meteorological Center made an accurate forecast that the rainstorm in the upstream in Wuhan would soon come to an end. Therefore, the State Council made a decision not to blow up the embankment, and the country was prevented from suffering from serious economic losses and the people from being displaced and becoming homeless.

In September 1994, Typhoon No. 17 formed in the South Pacific and made landfall in Wenzhou, Zhejiang Province. As the "Yinhe-2" had correctly predicted its intensity and specific landfall location, the department in charge in Zhejiang urgently organized people to reinforce buildings, transfer assets and valuable resources, and evacuate residents in an effort to reduce the loses caused by the typhoon as much as possible.

The "Yinhe-2" also made outstanding contributions in the fights against floods in the region of Dongting Lake in 1996 and in the Yangtze River Valley in 1998.

IV. Launching a Rocket in the Laboratory

Mathematics is a golden key to open the door of science. Theoretically, all scientific issues, and particularly technical engineering problems, can be solved through mathematical modelling.

Many of the scientific activities of mankind, especially various dynamic continuous systems such as aircraft flight, nuclear reactions, power grids and other dynamic physical tests, are so complex in term of their processes that it is difficult to organize them, as significant numbers of people and investments would be involved. In addition, there are also risks.

As a result, scientists came up with the idea that with the help of mathematical modelling, the processes of various scientific activities could be transformed into programs that can be operated in the computer and can simulate the process of a rocket launching, a nuclear reaction and aircraft flight. Is it not more economical, time efficient, effortless, secure and reliable?

This is computer simulation. It has a promising prospect in applications in the field of weapon systems, particularly in the design, test and systems check of the aerospace craft system, as well as in the fields of the aviation industry, nuclear power station reactor control, power gird control and others. Therefore, computer simulation technology has been the favourite of scientists ever since its debut, and has been longed for and pursued by developed countries.

However, to achieve a real-time computer simulation by keeping the simulation process in line with the actual physical process, an operation speed at least 100 megaflops is needed. Although simulation technology was invented by scientists as early as in the 1960s and was applied to missile design, computers at that time only operated at a speed of 10 kiloflops to one megaflops. Thus, early simulators were not digital simulators, but just analogue machines.

The mid- and late 1970s had seen a rapid development in digital technologies. The invention of the supercomputer "Cray-1", in particular, marked the advent of the "Computer Age" in which computers could run at a speed of 100 megaflops. Adopting the digital technology of the 100 megaflops computer, Analog Devices, Inc. (ADI) in the United States took the initiative of launching the "AD-10" simulator capable of advanced simulation languages in 1978. Subsequently, digital simulators in Western countries came thick and fast and were applied to military and civil economic developments in succession. Many simulation systems were established one after another, and played an important role in automobile manufacturing, petrochemical, energy and nuclear power plants and in aircraft, aerospace and weaponry research and development.

At that time, as a series of large scientific and technological engineering tasks represented by the "Nuclear Bomb, Missile and Satellite" programme in China were about to be developed at a much higher level, the need for the computer simulation technology to advance and support these tasks was quite urgent.

Shouldering heavy engineering modelling tasks, the former Ministry of Space Industry was longing to import a set of "AD-10" simulators from abroad, but soon abandoned such an idea. As early as over a decade ago, they had had experience in introducing a large simulator. At the negotiating table, in addition to imposing a high price, the most advanced model was the last thing the seller wanted to sell. Delays in delivery were a matter of course in spite of the signed contract, and when it was inappropriate to delay any longer, there would finally come an unacceptable condition: the machine could not be installed in Beijing, but in the Harbin Institute of Technology, which is thousands of kilometres away from the capital. This was tantamount to saying: "We want to place you out of reach of the machine."

The cruel reality struck the Chinese people once again. It would be only through Chinese efforts, rather than those of others, that China could have its own technologies.

On one night in December 1981, Secretary of the Science and Technology Committee Office of COSTIND Li Zhuang paid Hu Shouren a visit and said to him: "Simulation operation or semi-experimental simulation is necessary for the theoretical research, project verification, component or system testing and experiments of model tasks undertaken by the Chinese aviation and space departments. Although the analogue machine is fast in operation, it is low in accuracy, and is vulnerable to the outside environment. Thus, it is troublesome to use the analogue machine instead. But it is even hard to purchase such a machine from overseas. This has severely affected the development of China's aviation and space industry."

Staring at Hu Shouren with expectation, Li Zhuang said: "Can you help the nation out by sharing this burden?"

Hu Shouren froze at Li Zhuang's words. Could he spare some effort and researchers for the development of a digital simulator? The research and development of the "Yinhe-1" was at a critical moment, and everyone was working overtime for a breakthrough. In addition, they did not know any more than a layman about simulator technology.

Li Zhuang continued, saying, "You have worked on supercomputers for several years and have accumulated profound experience in technology. There is greater likelihood of success if you take charge of this task, and then we can be relieved. We have no such confidence if this task is undertaken by others."

On hearing this, Hu Shouren agreed immediately and without doubt, saying, "Alright, we will do this job!"

In the minds of the people working on Yinhe technology, responsibility is always of paramount significance. They shoulder the heavy burden with a disdainful attitude toward difficulties.

At the beginning of 1982, Hu Shouren soon led a group to Beijing and Shanghai for a thorough investigation of the demands and to perform a collection of information from among the users. Through further profound analyses and discussions and repeated verification of the research, they at last put forward a program for the research and development of an all-digital simulation computer.

In June, COSTIND held a hearing on the digital simulation computer program. Through thorough and comprehensive discussions, leaders, user representatives and technical experts thought that the research and development program of the NUDT was of innovative significance, and decided to entrust the crew from NUDT with the task of developing the first digital simulator in China.

In December, COSTIND hosted another hearing on the digital simulator programme, in which the research and development programme known as "Simulation main control computer plus simulation special-purpose computer" prepared by NUDT was reviewed. "Simulation main control computer plus simulation special-purpose computer" referred to the simulation system, consisting of a control processor, and a special-purpose simulator, in which the user would control the system through operating the main control computer.

In 1983, the research and development of China's first digital simulator commenced. The Institute of Computing Technology of

NUDT took charge of the research and development of the hardware system and the Department of Automatic Control of NUDT was responsible for the software. At that time, great progress had been achieved in the research and development of the "Y inhe-1" project, which was already on the stage of system assessment and trial calculations. In an effort to reinforce the major research and development force, the institute appointed Zhou Xingming, who was quite experienced in the research and development of supercomputers, as the team leader.

In April 1984, Hu Shouren, who had been working on the forefront of computer technologies for more than 20 years and was about to retire from his position as the Deputy Director of the Department of Computer Technology (and also the vice president of the Institute), volunteered to resign from the post as the project head and recommended Zhou Xingming as his successor.

In order to keep abreast with the world's advanced technologies, Zhou Xingming led the group to research the "AD-10", the most advanced mainstream model in the world at that time, and carried out all-around re-innovations by making use of its the leading design concept.

It is said that a thorough knowledge of yourself and your

enemies makes you invincible.

The team first carried out a performance analysis of their target computer, the “AD-10”. Due to a strict blockade of technology by the West, it was impossible to introduce a prototype from abroad and so a set of photos and some pieces of information from international journals were the only resources to which they had access. However, relying on years of experience in computer research and development and on their outstanding wisdom, they finally deciphered the secrets of the “AD-10” after two months of hard work. They discovered that its super simulation capability was mainly attributed to the following factors. First, it had a main-auxiliary computer structure, in which the main computer was applied to simulation programming and simulation non-real-time regulation while the auxiliary computer assumed the function of a simulation special-purpose computer, a sheer program performer devoid of operating system and programming language. Second, the auxiliary computer was designed with a fixed-point operation mode at only 16 bits, which helped to accelerate the speed. Third, the simulation special-purpose computer was of a heterogeneous multiprocessor bus structure, which was capable of processing different operations separately according to their types and was conducive to increasing the speed. Finally, it had a specially designed input and output interface processor, which could be used to connect the main computer and the simulated object.

Meanwhile, they also became aware of a major challenge in the research and development programme: the requirement that their machine be completely compatible with the world's mainstream digital simulator, the "AD-10". This was difficult, particularly in a condition without a reference system. Although relevant departments had tried with all their might to acquire a computer of this type, none was obtained for the enduring reason that it would be a violation of the technical export prohibition set by the CoCom.

As a result, they had to hunt for relevant materials and information by consulting substantial international bibliographies. Through profound analysis and verification, and with strenuous effort, they finally mastered the design concept of the model of the leading simulator at that time. On this basis, breakthroughs were achieved in a series of key technologies of digital simulators.

In 1985, the research and development of the "Yinhe Simulation-1" system was successfully completed.

According to the National Accreditation Board: "The 'Yinhe Simulation-1' is advanced in technological standards and is equipped with software that are compatible with hardware instructions and advanced language texts of the 'AD-10' of America, and its major technological standards have achieved and even surpassed that of the 'AD-10'. With a leading technological advantage of the 1980s, it has complemented China's deficiency in the field of digital simulators and is a major scientific and technological achievement of significance." "The 'Yinhe Minisupercomputer' is a self-developed high-performance minisupercomputer of China, which is stable and reliable in system...and is of the leading level in the early 1980s. Its successful invention has had important significance in advancing China's computer technological level as well as its economic performance, has narrowed the gap between China and other advanced industrial countries and is a major scientific and technological achievement of significance in China's computer industry."

On 6 June 1986, the "Yinhe Simulation-1" was put on display in Beijing. Dozens of leaders from the CPC, the government and the army paid visit to the exhibition and gave high evaluation marks to the "Yinhe Simulation-1".

On hearing this news, the president of the manufacturing company of the 'AD-10' in the United States came to China in person for an investigation. After a careful and detailed observation and study of the "Yinhe Simulation-1", he could not help complimenting its developers on the "Yinhe Simulation-1": "Your digital simulation technology is OK!"

America played the same old trick again after this compliment by revoking its prohibition on exports and flooding the Chinese market with its own products by cutting the price by 60%.

However, this trick did not work this time. The reason was quite simple: the "Yinhe Simulation-1" was cost-effective, as it was not only cheap in price, but also excellent in quality with a high performance.

In total, 11 sets of the "Yinhe Simulation-1" were manufactured and sold. These were installed by the army, institutions of higher learning and scientific research institutes. In addition, several simulation systems with the "Yinhe Simulation-1" as the focus were built up around the country, and were widely applied to simulation experiments of major projects in aviation, space, navigation and nuclear physics.

In a letter of thanks from the Second Institute of the Ministry of Aerospace Industry to the Institute of Computing Technology of NUDT, the Ministry said, "During the process of the research and development of the 'Long March 2' strap-on rocket, we have carried out pure mathematic simulations over 200 times and semi-physical simulations seven times and a half with the help of the 'Yinhe Simulation-1'. It has really made an outstanding contribution to the successful launching of the rocket when it was first launched."

Just in this project alone, the "Yinhe Simulation-1" had saved tens of millions RMB for the country!

Due to this achievement, COSTIND spoke highly of the "Yinhe Simulation-1" and praised it, saying, "It is one of the major simulation system equipment in contemporary China, and its performance is equivalent to that of the American special-purpose simulator 'AD-10, but it is cheaper than the 'AD-10' with helpful home-based tech support, all of which has earned it a high reputation among users. Due to the invention of the 'Yinhe Simulation-1', it is probable that China will develop high-level simulation systems and carry out many experiments that could not be done in the past..."

Because of its outstanding performance, reliable quality and excellent results manifested in the application, the "Yinhe Simulation-1" has made digital simulation, in particular semi-physical simulation, an indispensable means to shorten the development lifecycle of massive project model tasks, and to improve their development quality and save their development costs, thus shattering people's blind faith in imported machines.

As pointed out by COSTIND in a briefing, "Currently, many old users as well as new users have wiped out their old ideas of importing a simulator from abroad and are hoping that NUDT will invent the second-generation simulator as early as possible."

According to China's long-term development plan, there would be several heavily tasked key projects in the sector of national defence and military during the Eighth Five-Year Plan period and the Ninth Five-Year Plan period. For the real-time simulation function of some complex systems among those projects, the "Yinhe Simulation-1" was far from meeting the demand. To complete such project tasks as mentioned above, either several simulators must work together, or a new simulator must be developed of a higher performance than the current first-class simulator in the world.

Under this circumstance, COSTIND put the research and development of the "Yinhe Simulation-2" on the agenda shortly after the "Yinhe-2" project had been launched.

In October 1989, Deputy Director of the Commission of Science and Technology of COSTIND Nie Li hosted a reporting conference on the comprehensive verification and examination of the "Yinhe Simulation-2".

Ding Henggao, the Director of COSTIND, explicitly pointed out that "The research and development of the 'Yinhe Simulation-2' must be done in a way through which the favourable situation of China's 'reform and opening-up' policy should be brought into full play and the principle of 'introduce, digest, absorb and innovate' should be abided by in a serious manner. We must be brave enough to leap forward and avoid repeating carrying out the research and development at low levels."

When it came to the model to be developed, Director of the Commission of Science and Technology of Aviation Industry Corporation of China Chen Huaijin, an important user of the digital simulator, said, "Currently there are two major directions for the development of simulators, i.e., the development of heterogeneous multiprocessors and the development of homogeneous multiprocessors. To adapt simulation experiments of different types on different scales, it is necessary to develop both of them."

According to experts at the meeting, the heterogeneous multiprocessor is at a special advantage in the real-time or super-real-time simulation of continuous dynamic systems, and particularly in the simulation of physical circuits; the heterogeneous simulator, like the "System 100" recently released by ADI in America, is the most typical representative of the leading mainstream simulator in the world, the simulation capability of which is as high as two to four times that of the "Yinhe Simulation-1".

More than 30 simulation and computer experts at this meeting unanimously agreed on the comprehensive verification report made by Professor Jin Shiyao from the Institute of Computing Technology of NUDT, and had formed three decisions.

First, in an effort to takeover and surpass the world's advanced level, the research and development of the "Yinhe Simulation-2" must be finished in a short period;

Second, the research and development of the "Yinhe Simulation-2" should make China capable of completely getting rid of its inferiority in the field of digital simulation.

And third, strenuous efforts must be made to make the "Yinhe Simulation-2" competitive in the international market.

In November 1989, just one month after the hearing, COSTIND approved the research and development task of the "Yinhe Simulation-2". It was stated in the written instruction by COSTIND that the Institute of Computing Technology of NUDT and the Department of Automatic Control of NUDT would be responsible for the research and development project, with the Institute of

Computing Technology in charge of the development of the hardware and the Department of Automatic Control responsible for the development of an advanced simulation language. Jin Shiyao was appointed as the chief designer, and Lu Xicheng, Zhou Qihong and Dai Jinhai as the deputy chief designers.

On that same day, the leaders of COSTIND and the Institute met with Jin Shiyao in the conference room and entrusted him with the task.

A leader said, "Professor Jin, you are going to undertake the research and development task of Yinhe Mini-supercomputer. It is a digital simulation main control computer, a front terminal of the 'Yinhe-1', and also a prototype for the military."

Jin Shiyao replied with a resonant voice, "Yes. We promise to complete our task!"

The leader continued, saying, "The computer must be 100% compatible with VAX11/780. It should be available for a direct operation of the program on this computer without any recoding."

Jin Shiyao was stunned at these words. To develop a machine that is identical, and even better, in performance to another one which could not be obtained as the research reference—would it be possible? Is it not an impossible mission, designed on purpose?

Standing up straight away, however, he said, "I'll take the order! Even if obstacles ahead are as harsh as wading into hot oil, I'll overcome all the difficulties and fish a compatible computer out from the hot oil!"

The leader laughed, saying, "Your loud answer indicates that you are quite determined. Well done. From now on, we are counting on you for the development of the simulator!"

However, Jin Shiyao also pleaded, "Please find us a sample computer if you possibly can."

On the train back to Changsha, a departmental leader of COSTIND said to him, "Do not have blind faith in sample machines. Several national departments have already carried out research by following suit, but they all failed. You should learn from this lesson, and follow the path of self-innovation."

Jin Shiyao did not repeat the failure of others. In line with the international leading design concept, he carried out innovations and successfully achieved breakthroughs in key technologies like microcode. While digesting and developing the DEC database, he independently developed a Chinese-English office administration system. Taking advantage of the current network functions of VAX, he developed the miniature converter data packet assembler/

disassembler (PAD), and with his help, it was possible to manufacture a power device in China that was interchangeable with those from abroad. He realized the direct operation of all software of VAX-11/780 in the "Yinhe Minisupercomputer" without further modification, plug-ins that he developed were physically compatible with and capable of substituting imported plug-ins and he successfully mastered key research and development goals, including the key technology of drive programming for the equipment of the DEC. As a result, he also successfully developed the "Yinhe Minisupercomputer" for the country, a computer that both had the international leading advantage and featured Yinhe technologies.

According to the National Accreditation Board, it was an achievement coming under the guidance of the "introduce, digest, develop and innovate" policy of the central government. During the process of research and development, on the basis of the research experience from developing the 100 megaflops Yinhe computer and at the same time when the physical-level compatibility research and development was in progress, technologies that have Chinese characteristics are formed through innovation and by taking into consideration China's current level of design and manufacturing. This means a wonderful step forward in the nationalization of technologies and the development of technological applications.

In total, six sets of the “Yinhe Minisupercomputer” were manufactured. They were used as the front terminal of supercomputers, as the main control computer of simulation systems and as host processors at computing centres in various fields, including nuclear physics data processing, space flight simulation, seismic data analysis, scientific research and teaching and management. They have made significant contributions to China’s defence and economic development.

One should start at the same level as others in order to climb higher than they do. With scores of research and development personnel under the leadership of Jin Shiyao and the Chief Designer Group, the crew had spent over half a year on reading and consulting substantial materials and information about the “AD-100”, the most advanced machine in the world, to know in detail its design concept and technological advantages, and had spotted more than 200 mistakes about it in the relevant technical materials.

Making good use of its advantages and avoiding weaknesses on this basis, they had been moving bravely forward. They elaborately designed a technical scheme for the simulation system of the synchronous-bus heterogeneous multiprocessor with a floating-point word size of 65 bits. According to the performance assessment predication, machines manufactured in line with this scheme were completely capable of matching international mainstream simulators.

To surpass the world's leading technological level, the ability to make technologies a reality was also in need. The plug-in board of the "Yinhe Simulation-2" was twice as dense as that of the "Yinhe-2", and was assembled in a dual-board structure, with a mother board and a smaller one. On the two sides of the board, although they were only as big as a regular magazine, there were as many as 14 layers of assembled circuit boards covering over 500 components and thousands of connecting lines. Thus, it was very difficult to manufacture this kind of plug-in board. However, under the leadership of deputy designer of the CAD system Lu Xianzhao, the team upgraded and improved the CAD program of the "Yinhe-2", and then conquered the technical difficulties in high-density assembly by successfully inventing an excellent wiring design tool.

In order to fulfil the design and manufacturing task for the very-large-scale integration (VLSI), a young instructor named Wang Mingshi went abroad alone and engaged in the work of logic design in a foreign country. He diligently learned new concepts, measures and tools about design, achieved breakthroughs in techniques and completed tasks on time with a high quality, ensuring that VLSI could be successfully put into manufacturing each time.

In face of obstacles in the test of VLSI, Zhang Deqing, a senior technician, and his colleagues independently developed an integrated circuit testing platform after a year of hard work. The platform could be operated at a frequency as high as 90 MHz and was capable of testing more than 300 pins. This not only ensured the progress of the project, but also saved the nation over one million RMB in expenses and costs.

In an effort to ensure a global leading advantage of the simulation software, over 20 comrades working on the software system had spent two consecutive years developing six sets of large-scale comprehensive software programs with a total of 250,000 lines, which included the compiling program for the advanced simulation language, the advanced human-computer interaction program and the real-time I/O assembler program. They had also achieved a series of new innovations, like the parallel processing and the multiple iterative loop algorithm. With these programs and innovations, the structural advantage of the multiprocessor was exerted to the full to help the "Yinhe Simulation-2" achieve a better performance in practical use.

On 24 March 1993, the Working Conference of the Trial Calculation of the "Yinhe Simulation-2" was held in the NUDT. At the conference, 12 scientific and research units and institutes in the field of space and national defence science, making use of the intervals between their trainings, finished the real-time or super-real-time simulation calculation of two sets of large-scale scientific engineering subjects. Comrades from the space sector, in particular, had successfully completed the programming for a major national sophisticated technique project just by making use of the time in three noon shifts and three short night shifts. Satisfactory results were also obtained after data of such high-tech projects as the "Long March 2" strap-on rocket had been input into the simulation system. A series of trial calculation results had shown the following results:

The "Yinhe Simulation-2" was four times better than the "Yinhe Simulation-1" in terms of speed acceleration.

Due to the "Yinhe Simulation-2"'s excellent compatibility, the "AD-100"'s simulation software could be directly operated on the "Yinhe Simulation-2", with its operating speed, capacity, user interface, simulation capabilities, software environment and performance-price ratio all better than that of the "AD-100".

The "Yinhe Simulation-2" was capable of meeting the

simulation demands of new weapons to be developed during China's Eighth and Ninth Five-Year Plan periods, and could be directly applied in the field of aerospace, aviation, automobile, energy and chemicals. It was of important strategic significance in advancing the application of simulation technologies in the modernization drive of the economy and national defence, as well as in promoting the research and development level of computers in China.

Six of the 12 participating units and institutes immediately expressed their desire to purchase the "Yinhe Simulation-2".

On 17 May, the evaluation test for the "Yinhe Simulation-2" commenced. During the 72 consecutive hours of the evaluation, the machine experienced no stoppage or fault and operated smoothly until the final manual shutdown.

On 22 June 1993, experts of the National Accreditation Board who came from various sectors and fields in the nation stared avidly at the terminal of the "Yinhe Simulation-2", the screen of which showed in succession that:

The operation speed reached 87.5 million lines of code per second.

Two styles of floating-point word size of 65 bits and 53 bits respectively were achieved in operation accuracy.

There were as many as 256 channels for the real-time I/O function.

These data were tantamount to an announcement to the world that the "Yinhe Simulation-2" was one to two times better in terms of its comprehensive capabilities than the world's most advanced simulation computer that could be purchased in the international market.

This was an international leading level that was worthy of fame.

In January 1994, recommended by 156 academics and over 300 experts, the successful invention of the "Yinhe Simulation-2" was listed as one of the top 10 pieces of science news in 1993.

In 1995, the "Yinhe Simulation-2" won the National Science and Technology Progress Award. Jin Shiyao, Zhou Qihong and Wang Purong were awarded the second-class merit.

In September 1995, a piece of good news came from the Beijing Simulation Center: "The whole simulation process of the Long March 3 strap-on rocket, which is the largest in scale and the most complicated in history, has been finished. The simulation results are accurate with a high quality. The period of simulation design is 80 days before it has successful passed the examination and approval. The research and development lifecycle has been shortened by nine months. The 'Yinhe Simulation-2' has made huge contributions to China's development as well as to the development of the national defence."

During the mid- and late 1980s and the early 1990s, the Institute of Computing Technology of NUDT had, at the same time when it was assuming large-scale important tasks, also taken charge of a series of projects associated with the national research program known as the "863 Program", the National Natural Science Foundation of China, the National Defence Pre-Research Foundation Program.

The Computer Software Teaching and Research Office and the Artificial Intelligence Computer Teaching and Research Office of ICT and the Software Laboratory of Wuhan University co-developed the object-oriented integrated software development environment "GWOOSE";

The Computer Aided Design and Aided Measurement Teaching and Research Office of ICT completed a project titled "YC-2000 CAD System for High-speed Large-scale Electronic Equipment";

The Artificial Intelligence Computer Teaching and Research Office of ICT developed the military expert system "Sanjie 3";

The Computer Software Teaching and Research Office of ICT

finished the non-monotonic reasoning system "GKD-NMRS";

Finally, the Artificial Intelligence Computer Teaching and Research Office of ICT has developed the large-scale neural network simulation system "GKD-N2S2";

In the early 1990s, China's Central Military Commission conferred the honorary title of the "Pioneer of Science and Technology" to the Institute of Computing Technology of NUDT.

Chapter V
Young and Vigorous Strategic Weapons in a New Era

The intensified strategic weapons competition transformed the computer field into a new arena of competition for the world's major countries.

In this context, the Institute of Computer Science and Technology of NUDT successfully developed the "Yinhe-3", which is a 10-gigaflops supercomputer, achieving a leap from multi-processing (MP) to massive parallel processing (MPP). Such an achievement is also a "phoenix nirvana" for the Yinhe researching and developing team.

I. The Launching of the 10-gigaflops Supercomputer Project

In order to permanently occupy the commanding heights of the world's strategic weapons and technology, the United States government proposed a plan titled the "Accelerated Strategic Computing Initiative", aiming to increase the computing speed of high-performance computers in parallel with the development of United States preparedness plans and demands. The United States government invested billions of dollars to develop high-performance computer systems for weapons-oriented strategic experiments and planned to launch 10-teraflops supercomputers in 1999, 30-teraflops supercomputers in 2002, 100-teraflops supercomputers in 2004, and to achieve petaflops-capable supercomputers in 2007. The United States government continued to push forward high-performance computing technologies to new heights and has become the leading country in world computer technology. Since the creation of the TOP500 supercomputer rankings, the first top position has been generally occupied by American machines, and most of the top 10 machines are made in the United States.

The United States made use of continuously launched supercomputers to perform strategic weapon simulation experiments and realized the upgrading of strategic weapons technology. It also utilized its unique supercomputer technology to develop and deploy its national missile defence systems and theatre missile defence systems. Even some territories and territorial seas of other countries were included in the defensive scope of the United States.

In the 21st century, the development level of a country's supercomputer, in a certain sense, represented its national security level.

High-performance computing has become the basis on which a country can maintain its national security!

The world supercomputer technology competition has become more and more intensified. In this new competition, China must not lose again.

Thus, in the context of China implementing contracting reforms for scientific research funds (such that scientific research projects require user investments), how should China start and progress the development of national supercomputers?

The establishment of the "Yinhe-2" project had been delayed five years because it was difficult to fulfil the needs of its users. Five years was actually the lead time for developing a whole new generation of supercomputers. In addition to the traditional users of the aerospace industry, meteorology and seismic data processing, the world's supercomputers at that time were also applied by some new users, such as those performing encryption and deciphering studies, military operations and battlefield simulations, VLSI design and demonstration, bioscience and drug research.

The strategic core status of supercomputers had certain contradictions with the high investment, high risk, imminent application and huge difficulty to fulfil the needs of its users. How can this contradiction be solved?

Let us look at how the United States solved this problem. We will first talk about the investors in different generations of computers in the United States

The investors for the first computer to really realize scientific computing was the United States Air Force.

The investor for the first all-transistor computer "TRADIC" was the United States Department of Energy, which specializes in strategic weapons research and management.

The research and development funds for the 100-kiloflops supercomputers, known as "Stretch" and launched in 1956, was also invested by the United States Department of Energy.

The investor in the "Cray" series of supercomputers, developed under the auspices of Cray, the father of the world's supercomputers, was Los Alamos Laboratory.

Supercomputers such as the "Titans" and "Sequoia" could compete with Chinese supercomputers in the previous years, and they once occupied the top spot of the international TOP500. These supercomputers were also funded by the United States Department of Energy. All in all, all these research and development funds for each generation of supercomputers in the United States were government investments, and they were really a huge investment.

The reason why the United States government was so generous to computer industry is that they believed that national security, aerospace and aviation industry as well as meteorology were fields that must be invested in. Without high-level computers in these areas, it would be difficult for the United States to move forward.

Based on the same recognition, the leaders of COSTIND gave important instructions after receiving the report about the development status of the Yinhe supercomputer, saying "The Yinhe supercomputer must find a 'required investment field' such as weather forecasting, fluid mechanics calculations, nuclear power engineering data processing, etc. Through the long-term cooperation from both parties, we can continue to promote the sustainable development of the Yinhe business."

For this reason, COSTIND decided in March 1994 to launch the "Yinhe-3" 10-gigaflops supercomputer project. The Yinhe New Technology Development Company, since then representing a "required investment field", signed a project development contract with the Institute of Computer Science and Technology of NUDT.

Who would then be in charge of the "Yinhe-3" project? Everyone looked to Lu Xicheng.

This expectation was a matter of course. Lu Xicheng was 48 years old. As a scientist, he was in the prime of his life. He was the head of the computer department and director of the Institute, which was convenient for him to carry out overall planning and coordination. He was also a prominent representative of the third generation of Yinhe researchers. In the 1970s, he had participated in the development of a national key project, the "Central Computer of the 'Yuanwang' Surveying Ship". He went to the Pacific twice to perform his tasks and both trips were successful. In the early 1980s, he was sent by the country to study in the United States. He was one of the earliest experts studying network technology in China. After returning to China, he joined the research and development team of the "Yinhe-2" supercomputer and took the initiative in researching and developing the world's most advanced high-speed computer network technology at that time. The high-speed network software system developed by his team enabled the "Yinhe" supercomputers to support online supercomputing. He was the deputy chief designer of the "Yinhe Simulation-2" computer, responsible for developing software. He made significant contributions to achieving the real-time all-digital semi-physical simulation, which was urgently needed by the national aerospace and aviation industry at that time. Since then, he had led his team and successfully developed the core router of the "Yinhe Yuheng 9108", effectively promoting the progress of key technologies for key equipment in China's network. He led his team to develop the "8100" high-speed network real-time security monitoring system, which made important contributions to the maintenance of national and military internet security. He won four first prizes, one second prize, and multiple third prizes at the National Science and Technology Progress Awards. Moreover, he was awarded a First-Class Merit Citation one time, and won the Major Technical Contribution Award for the Army and the Science and Technology Awards of the Ho Leung Ho Lee Foundation.

Lu Xicheng also thought about who the best candidate would be to lead the “Yinhe-3” project. However, he did not think of himself. This candidate was not only critically related to the development of the “Yinhe-3”, but also concerned about the long-term development of the entire Yinhe research team and the long-term development of the Yinhe business.

Computer research and development is the realm of young people. Alan Turing was only 24 years old when he founded computer theory in 1936. In 1946, when the Americans Presper Eckert and John Mauchly successfully developed the first computer in the world, they were just young people who had left school not long before. Von Neumann, known as the “Father of Computers”, was only 42 years old when he developed the famous “Von Neumann principle”, so it’s no surprise that the backbone of individuals who developed the "Yinhe-1" and "Yinhe-2" were basically university graduates in the 1950s and 1960s. They made significant contributions to China's supercomputing business in terms of long-term scientific research. Now they had reached or were starting to reach retirement age. At the same time, the new generation of computer talents who had grown up after the implementation of the "reform and opening up" policy not only had rich knowledge and profound science and technology expertise, but had also accumulated rich experience in scientific research through their work on the "Yinhe-1" and "Yinhe-2". They were the future of the "Yinhe" business, and the business needed them to stand out in this challenge as quickly as possible.

Lu Xicheng made up his mind that they would not only achieve a new leap in Chinese supercomputer technology through the "Yinhe-3" project, but also would achieve the "phoenix nirvana" of Yinhe team, in order to build a young team for the long-term development of the national supercomputing industry and to ensure that there would be adequate and competent successors for Yinhe, thus always retaining the youthful vitality of the team.

However, the leader of this young and vigorous Yinhe team could not be anyone else but Yang Xuejun.

Yang Xuejun was a young expert who grew up around the

Yinhe projects. In the early 1980s, after graduating from Nanjing Communication Engineering College, he was admitted to the National University of Defense Technology and studied under the guidance of Professor Ci Yungui. At the same time, he participated in the late-stage development of the "Yinhe-1". He even chose the research of the "Yinhe-1" compiler for his master's thesis. After graduating 1986, he and Tang Yuhua (his wife) both devoted themselves to the development of the "Yinhe-2". Soon after that, he enrolled himself in a doctoral program under the guidance of Professor Chen Fujie, the leader of the "Yinhe- 2" project. He selected a parallel processing technology closely related to the "Yinhe-2" as the topic of his doctoral thesis. While he was working on his doctoral thesis, he also participated in the development of the "Yinhe-2". After reading Yang Xuejun's doctoral dissertation, Ci Yungui instructed, "We must by all means keep this comrade with our research team, so that he could continue researching and developing the 'Yinhe-2' project after graduation". After the completion of the "Yinhe-2" project, Yang Xuejun and another three comrades were sent abroad by the school to do pre-research for the "Yinhe-3" for seven months, during which they expanded their international vision, understood and mastered the world computer technology trends and developing trends. Such a young scholar with a solid academic background, broad academic vision and large-scale engineering practicing experience is undoubtedly worthy of the trust and expectations of the Yinhe project.

On that day, the university leaders collectively handed over the task to Yang Xuejun. The moment that Yang Xuejun realized that he was appointed as the chief designer of the "Yinhe-3", he became nervous. Opening his eyes wide, he waved his hands subconsciously, saying, "Chief designer? I'm afraid I cannot do it. I might be ok to be an assistant for the chief engineer to help and give some suggestions or advice."

The university leaders encouraged him, saying that "The university Party Committee made this decision after serious and careful consideration and deliberation. The Party Committee believes that you can do it. We believe you will do a good job and surely will open up a new world for Yinhe!"

With the approval of COSTIND, the Party Committee of NUDT established the "Yinhe-3" research and development team with Lu Xicheng as the chief commander, Yang Xuejun as the chief designer, Zou Peng as the deputy chief commander, and Zhang Min as the deputy chief designer. The average age of this team was only 36, and the average age of the 19 principal designers and deputy principal designers was 40. However, the deputy chief designer was only 40 years old and the chief designer Yang Xuejun was only 31 years old at that time.

In order transmit the experience of previous Yinhe generations, and to aid with transmitting ideology in order to help and assist the new Yinhe generations, the university Party Committee also established an advisory committee for the "Yinhe-3" project with Yang Xiaodong and Li Xiaomei as the major representatives.

Therefore, the research and development of the "Yinhe-3" supercomputer presented the spectacular scene of a splendid sunrise and beautiful sunset, with groups of stars lightening up the Milky Way under the wide and starry sky.

Practice has proven that these decisions of the Party Committee of NUDT were extremely far-sighted. This young team, with the help of the old Yinhe pilot team, quickly became the backbone of Yinhe, continuously promoting the development of our national supercomputer technology to achieve leapfrog development and ultimately to push forward Chinese supercomputer technology.

II. Adventure is an Opportunity to Achieve Leapfrog Development

In August 1968, the company Fairchild Semiconductor, which had invented the integrated circuit, had the problem of employees being not of one mind due to poor management. The top talents left the company one after another. Only three experts decided to stay, namely Robert Noyce, Gordon Moore and Arthur Rock. They had to seek a strategy to get the company out of trouble.

After some in-depth analysis, they believed that computer development depended entirely on the development of components and that the components mainly referred to integrated circuits. They also believed that the computer chip market was the most promising semiconductor market. Thus, they figured out that whoever developed and produced integrated circuits with highest integration level actually could dominate this market. Therefore, they decided that Fairchild Semiconductor should focus their strength on this aspect and go all out to develop integrated circuits.

This conclusion at first was only a strategy of Fairchild Semiconductor to get the company out of trouble. However, the facts later showed that it was actually the beginning of a major revolution in the history of computer development.

After they determined the development direction of their company, they set up the Intel Corporation that specialized in developing and producing integrated circuit products. Three years later, the company introduced the "4004" microprocessor, which integrated 2,300 transistors on a 3 mm by 4 mm chip, concentrating a microprogram on a single chip. It not only helped to develop desktop computers, but also facilitated the emergence of laptop computers. In addition, the microprocessor was widely used in automobiles, airplanes, ships, satellites, and even some small products such as watches, cameras, telephones, televisions, and toys. By that time, human society had really stepped into the information era.

In the late 1980s and early 1990s, 64-bit high-performance

microprocessors were introduced to the world. Supercomputers of 100 megaflops and of one gigaflops were easily installed into small-scale casings. In other words, the scientific calculations that could only be performed in the past by giant supercomputers such as the "Yinhe-1" and "Yinhe-2" could now be completed by a small personal computer.

The 64-bit high-performance microprocessors, although having a certain impact on the supercomputer market, still could not replace supercomputers. On the contrary, this major technological breakthrough brought great advantages to the development of supercomputer technology. It helped realized the great dream of developing MPP supercomputers to solve large-scale scientific and engineering computing problems and large-scale data processing based on a design with microprocessor chips.

However, there were two major defects in this dream. On one hand, the theoretical peak computing speed of a MPP architecture supercomputer was still distinctly different than the actual performance that users could achieve. The service efficiency of this architecture was very low, usually less than 30%. On the other hand, the programming models on the MPP computer and a computer with traditional architecture were totally different, and therefore programming would be difficult for users. There were many technical challenges ahead. These challenges included the Internet bandwidth and latency between nodes, the efficiency of synchronization mechanisms, the equalization of computing and I/O capabilities, the capacity and bandwidth of storage units, the support environment for programming with MPP architecture, large-scale parallel processing methods, the differences between local storage access and remote storage access and the smooth scaling capability of the system.

In a word, the MPP supercomputer was difficult to develop and produce, but it would be even harder to use!

These two sources of difficulties caused many large international companies to suffer serious setbacks while exploring MPP technologies. Some companies even closed down. As a result, many companies felt depressed and frustrated about MPP. At that time, the exploration of MPP in China had also just started.

In contrast, the Yinhe research team were more familiar with and skilled at using MP, which they had already used in the development of the "Yinhe-1" and "Yinhe-2". Therefore, it was a likely means to achieve a leap from one gigaflops to 10 gigaflops. However, it would be impossible to use this architecture to achieve a leap to 100 gigaflops or even higher levels.

Thus, the question was as follows: would the "Yinhe-3" continue to make use of the MP architecture which engineers already knew well and managed with ease? Or it would take the new path of MPP, which was full of risks and adventures and barely explored?

The chief engineering team resolutely made three major decisions after extensive consultations with experts, based on support for both ideas mentioned above.

First of all, they would aim at 100 gigaflops technology and get all technical preparation work ready for subsequent leapfrogging while achieving the leap development of 10 gigaflops.

Secondly, they would aim at the future mainstream technology MPP in order to ensure the high quality of the "Yinhe-3" project, with a high starting point.

Finally, they would target the demands and requirements of applications for large-scale scientific and engineering computing and large-scale data processing. They would spare no effort to break through the key technologies of the MPP supercomputer system and develop efficient and easy-to-use machines for users.

The team leaders explained the main reasons for their decisions: Although development of MPP is risky, risk is an opportunity to surpass the competition. Only when we are brave enough to meet the challenges, they said, can we achieve and surpass the world-class level as quickly as possible. Following the footprints of others is not our style!

In order to realize this large leap in the history of Chinese computer development, the team determined the guiding ideologies of "standing tall and seeing far" and "catering to the pleasure of others".

"Standing tall and seeing far" actually means overcoming the latest technologies and achieving lofty goals from a high starting point. For example, applying advanced CPUs in hardware design enabled them to break through the bottleneck regarding the NUMA structure of distributed shared memory, the high-efficiency synchronization mechanism and remote direct transmission of data with large, medium and small granularity. In addition, overcoming other key technologies, such as the 64-bit microkernel operating system, distributed parallel I/O, automatic parallel recognition, parallel debugging tools, the Asynchronous Transfer Mode (ATM) network and so on, would make sure that the "Yinhe-3" could meet the high-performance requirements for large-scale scientific and engineering calculations and large-scale data processing, good application environments and strong network computing capabilities.

"Catering to the pleasure of others" means designing the kinds of machine that users want and expect. The starting point and destination of their research work was based on the rule that users could afford the machine and could easily use it well. For example, they strived to integrate the "Yinhe-3" with international mainstream models by opening up their architectures and fully abiding by international standards. At the same time, they also focused on emphasizing the autonomous copyright of system software, breaking the monopoly of Western countries and mastering the software technology initiative. Moreover, they realized the system scaling function, which would ensure that users' needs were met by increasing or decreasing the number of CPUs, no matter whether they needed 100 megaflops, several hundred megaflops, gigaflops, several gigaflops, 100 gigaflops or even hundreds of gigaflops. Furthermore, the low-power design and CMOS technology were adopted to reduce the machine's power consumption. High-density assembly methods were used to reduce the size of the machine. Advanced cooling technologies were implemented to reduce the machine's impact on the environment. The research team also cooperated with the application units in order to develop the application systems at the same time, so that the computer could be put into use immediately.

COSTIND fully supported their bold initiatives.

As the MPP technology was still in the initial stage of exploration, the exploration results were limited. In addition, Western countries imposed a tight blockade on their technologies. As a result, the Yinhe research team only got very limited reference materials. In order to provide them with strong information support, COSTIND organized a three-day MPP expert seminar in April 1994. More than 10 experts from Tsinghua University, Fudan University, the 631 Institute, the 15 Institute and some other units shared whatever they had obtained from their own research, which expanded the horizons of Yinhe research personnel and greatly benefited them.

On 6 July, COSTIND further invited experts from these units to participate in the demonstration meeting of the "Yinhe-3" project. After the experts learned about the development plan from the research team, they believed that the plan had two world-class features, namely a world-class advanced level and world-class development difficulty.

The Yinhe research personnel decided to use extreme measures to solve the world-class difficulties.

There is a botanical garden in which birds twitter and one can smell the fragrance of flowers and green tress on the outskirts of the

NUDT in the southeast direction, more than 20 kilometres away from the school. The botanical garden was very quiet. The leading group of the "Yinhe-3" project and more than 20 scientific research personnel carried related documents, backpacks and their daily necessities, moving into a row of bungalows there and starting their closed research. They had no holidays and no family visits. They worked all day and night, and they only took short naps when their eyes could not stay open. After a short sleep, they woke up and continued their work. After three months of such hard work, they successfully completed the system design of the "Yinhe-3". They developed a system program of MPP architecture with single-machine processing nodes, three-dimensional ring network interconnections, and distributed shared main memory, initially achieving the technical leap development from MP to MPP.

The older generation of Yinhe researchers dedicated themselves to protecting the young people in their process of innovation. During that year, numerous computer development experts over 60 years old (such as Zhou Xingming, Chen Lijie, Yang Xiaodong, Li Xiaomei, Tan Zhengxin, Peng Xinjiong, Li Sikun, Huang Kexun and Su Changqing) passed on valuable experience, experimental methods and technical information to the young generations. Professor Zeng

Xiaode applied an important research achievement to the "Yinhe-3" project. This research achievement cost eight years of effort and was accredited to have reached the international advanced level in the tests of the National Natural Science Foundation of China. In the design process of the key technology of the ASIC chip, Zhou Qihong, Xiao Yingshi, Yang Chaoqun and other experienced veteran comrades volunteered to form an expert review group.

With the support of old experts, the new generation of Yinhe research personnel fought and battled firmly along the MPP road.

Ultra-large-scale integrated circuits are the heart of supercomputers. The design and development of a variety of dedicated chips was a key technology of the "Yinhe-3". Each of the variety of dedicated chips had hundreds of pages of design drawing, and each page was full of dense circuits. If there was a small design error for one circuit, it would lead to the failure of the entire chip. That is why many international companies had only achieved a 60% success rate for the tipping-out test of such chips. However, the "Yinhe-3" research team strived for excellence in their designs with strict management requirements. They did careful checks at all levels and reviewed again and again. In the end, they performed a very successful one-time chipping-out test.

In the 1990s, a new type of asynchronous transmission technology emerged internationally. In order to ensure that the "Yinhe-3" was abreast with the world's advanced technology, the research team boldly changed their design. It took two years to complete the design and development of new network technologies after overcoming a number of key Internet technologies. They continuously surpassed three major steps, from Ethernet technology to distributed optical fibre data network technology and then to asynchronous transmission, configuring the "Yinhe-3" successfully with the world's smoothest information highway.

If the operating systems of supercomputers are not compatible with the outside world, they will cause great inconvenience to users, and thus further affect the application market. Before the "Yinhe-3", China had always designed its own operating systems, but they could not be compatible with international popular system platforms. Therefore, they were determined to change this situation with the "Yinhe-3". They organized a team to forge ahead with courage and to thoroughly understand the most advanced Unix operating system technology of that time. Then they boldly carried out some innovations and improvements. Finally, they successfully developed the "Yinhe-3" system software, which had its own special features and was also in line with the world standards, thus enabling the "Yinhe-3" to meet the international standards.

During the implementation of the "Yinhe-3" project, heat dissipation was a key technology. If this problem was not solved well, this invaluable supercomputer could be burnt into a pile of scrap iron in seconds. After repeated explorations, the research team successfully solved a series of problems, making the heat dissipation of the "Yinhe-3" reach an advanced level in the world. In the past, it was necessary to build expensive dedicated air-conditioning systems for the supercomputers. Therefore, some people joked that "we can afford the supercomputer, but we cannot afford using the supercomputer." After the advent of this heat dissipating technology, ordinary air-conditioned computer rooms could meet the environmental requirements of the supercomputer, which greatly reduced the cost necessary to use and maintain it. This improvement was greatly welcomed by users.

The stability and reliability of the supercomputer is the primary concern for users. Whether in the aspects of the ASIC chip or its function plug-ins or its overall machine, the "Yinhe-3" fully implemented an advanced and perfect boundary-scanning mechanism. Combining two-level diagnostic systems that could support offline and online debugging, the "Yinhe-3" achieved accurate and fast fault location. The backup node configured by the system could flexibly replace the fault processor, and the system could be automatically reconfigured or downgraded. The reliability, availability, and maintainability of the whole system had reached the domestic leading level.

Chipset operating frequency was up to 100 MHz, which is already the advanced leading level in China.

The application environment was also at the leading level domestically.

The parallel algorithm was practical and efficient, and its comprehensive level was already at a leading level in China.

The engineering design was also at a leading level in China.

The visualization platform was the first such initiative in China.

This series of leading levels and first initiatives enabled the "Yinhe-3" iterations to successfully leap over one and surpass other technical obstacles in the history of Chinese supercomputer technology development:

It transitioned from MP to MPP in terms of architecture;

It evolved from small- and medium-scale integrated circuits to large-scale integrated circuits;

It moved from the complete purchase of chips to using independent designs of ASAC large-scale integrated chips;

It transitioned from a dedicated batch operating system to an open operating system Unix;

It expanded from an independent supercomputer system to a network computer system with supercomputers as their core;

It moved from flat logic design to hierarchical logic design;

It moved from a strictly required specialized computer room to a general working environment without special requirements;

It transitioned from a high-power supply to a low-power supply system.

The calculation speed of the "Yinhe-3" was 10 times faster than that of the "Yinhe-2", but its volume was only one-sixth that of the "Yinhe-2"!

At that time, the actual computing performance of MPP supercomputers was generally 20% to 30% of its peak computing performance, while the "Yinhe-3" reached 70%!

III. Dedication to Hong Kong Returning Home

In June 1997, Hong Kong, after being occupied by Western countries for a hundred years, returned back to motherland.

At the happy moment when Hong Kong was about to return to the embrace of the motherland, the "Yinhe-3", which had been carefully cultivated for three years by Yinhe research team, was finally ready to be shown to the public.

However, was the "Yinhe-3" strong and did it meet the requirements of the public? On 6 June, Chen Danhuai, the Deputy Director of the Department of Science and Technology of COSTIND, led seven experts to conduct a rigorous examination.

At 8:00 that evening, leaders, experts, and research personnel gathered around the "Yinhe-3". After seeing that it powered up and ran smoothly and stably, all of the people present acclaimed the project with uniform cheers and smiled with delighted relief.

At that time, the wind whistled, the lightning flashed and the thunder rumbled with heavy rain outside the computer room. The voltage of the computer room fluctuated severely. At one moment, it was up to 250 V, but the next moment it had dropped to 120 V. There was even a short-term power outage that evening.

Could the "Yinhe-3" survive and withstand such torment? The experts and leaders of COSTIND were nervous and worried immediately.

Chen Danhuai turned to the experts examining the machine for advice, asking "What should we do in this weather? There will be a great loss if the machine gets broken."

The experts examining the machine agreed that they should stop the machine first, wait and then restart after the thunderstorm had finished.

Chen Danhuai came to Lu Xicheng and Yang Xuejun and said, "The power voltage is too unstable at this moment. We recommend shutting the computer down first. How do you think we should do it?"

Lu Xicheng and Yang Xuejun looked very serious. As the chief commander and chief designer of the project, although they were fully aware of and confident about the stability of the "Yinhe-3," this was the first time they had encountered such bad weather since the testing of the "Yinhe-3", and so they were also concerned and worried about whether the "Yinhe-3" could survive such a storm.

However, the "Yinhe-3" was still running stably and in a good manner, like a steady reef under the turbulent winds and waves.

Therefore, Lu Xicheng and Yang Xuejun made a bold decision, saying, "Do not stop and continue to test machine!"

After 10 hours, 24 hours, 48 hours and 105 hours, the "Yinhe-3" was still running smoothly.

Finally, the experts examining the commander gave the instruction, "Stop the machine manually!"

The experts who examined the machine waited next to the machines in turn, and they were deeply impressed with the "Yinhe-3". "A computer that can resist such bad weather and run continuously and smoothly for four or five days—there is no doubt that its reliability and stability will be world-class!"

However, the physical examination of the "Yinhe-3" machine did not finish so quickly and easily. On the following day, 32 experts from all over the country were divided into three groups to test and evaluated the 11 special technologies of the "Yinhe-3". The experts sought hard to find potential defects. However, after they checked it out for three days straight, they still did not find any problems. In the end, they finally smiled and gave a high score to the "Yinhe-3".

The assessment of the "Yinhe-3" was ongoing. COSTIND required them to send the machine to its first user (a research institute in Beijing) for further assessment. The Yinhe research team knew that this was because the leading experts and users wanted to evaluate the anti-vibration and installation performance of the "Yinhe-3".

After two trips by express trains, a long-distance trip of nearly 20 hours, the "Yinhe-3" finally arrived in Beijing on the evening of 14 June. The research staff disregard the travel fatigue, started to open packages and assembled the computer at 7:00 p.m. after dinner. After finishing the assembly installation at 11:15 p.m., they closed the breaker and powered up the machine, and the "Yinhe-3" ran normally and smoothly.

The leader and the people of the Institute who came to visit were all surprised and astonished, saying, "In the past, it took a few days to assemble a supercomputer. But this 10 gigaflops supercomputer only took just a few hours. It was incredible."

On 19 June, the performance appraisal of the "Yinhe-3" officially kicked off. Ding Henggao, the Director of COSTIND, was personally present as the Chairman of the appraisal committee. Zhang Xuedong, the Deputy Director of COSTIND, Hu Qiheng, the Vice President of the Chinese Academy of Sciences, Zhang Xiaoxiang, the Chairman of the China Computer Federation and Wang Chengwei, a member of the Science and Technology Commission of COSTND, all took part in this appraisal activity as the Vice Chairmen of the appraisal committee.

A national appraisal committee composed of 38 famous experts (including 14 academics) conducted rigorous tests of the "Yinhe-3" system that included 128 operation processing nodes, eight backup processing nodes, eight I/O processing nodes and corresponding software systems.

More than 10 application units brought more than 20 testing questions, involving many fields such as chemistry, petroleum, meteorology, engineering physics, fluid mechanics and structural analysis. After inputting these questions into the machine, users and experts from all over the country kept silent and watched the screen attentively. With the soft sound of the computer while it was running and after continuous flickers on the screen, the data of the China

Earthquake Administration came out, and the data of the Ministry of the Petroleum Industry, the data of the Ministry of Aerospace, the data of the National Meteorological Center and the nuclear power experiment data all appeared on the screen one after another.

The calculation results of more than 20 questions were all correct!

"We made it! We made it!" The lobby of the computer room was full of cheers.

On behalf of the appraisal committee, the Director of COSTIND wrote solemnly on the "Yinhe-3" appraisal book:

A number of technologies of this supercomputer rank as the country's leading level in China, and its integrated technology has reached the current world-class advanced level. The test results of the relevant application units to the computer show that the "Yinhe-3" runs in a stable and reliable state with good scalability and high parallel speedup. It can achieve wide applications and a good high-performance-to-price ratio...The "Yinhe-3" supercomputer can be applied to large-scale scientific computing, large-scale data processing in many fields such as chemistry, petroleum, meteorology, engineering physics, fluid dynamics or structural analysis. At the same time, it will provide major boots to the development of China's national defense, economic construction, science and technology.

Zou Jiahua, the Vice Premier of the State Council, came to the appraisal meeting on that day and watched the on-site calculation demonstration of the "Yinhe-3". He also met with the main development personnel.

Vice Premier Zou Jiahua pointed out that with the rapid development of science and technology today, high-tech research work must be courageous to emancipate the mind and take the creative development path; meanwhile, we also need to pay great attention to the promotion and application of high-tech achievements, focusing on promoting products to the market and expanding the application areas, thus improving the efficiency of applications.

The success of the "Yinhe-3" symbolized the success of the new generation of the Yinhe research team, represented by chief commander Lu Xicheng and chief designer Yang Xuejun, demonstrating that they had successfully taken over the "relay baton" of Yinhe.

At this point, the three strategic objectives of the "Yinhe-3" project had been successfully fulfilled: mastering the key technologies for higher-performance supercomputers, developing high-performance machines that were highly popular with users, and successfully taking over the "relay baton" of Yinhe!

The "Yinhe-3" was named as one of the National Top Ten Scientific and Technological Progress Award winners of the year. It won the National Science and Technology Progress Award and the first prize of the ministerial-level scientific and technological awards.

The media commented, "The 'Yinhe-3' has prepared a liberal present to celebrate Hong Kong returning to the motherland!"

A Hong Kong media outlet reported that "Hong Kong will return to the embrace of the motherland soon. However, there are still people who doubt whether China can manage this international metropolis. Now, this question has been answered by the emergence of the 'Yinhe-3'. It should be noted that Hong Kong's former administrator, Britain, has not yet developed a 10 gigaflops supercomputer. We firmly believe that a country with an advanced technological level will not lose to others in the aspect of governing abilities."

IV. Weather Prediction System

A global consensus is that "the application of supercomputers is even more difficult than their development."

The reason lies in multiple insurmountable discipline gaps between the technology and applications of supercomputers. It is the gaps that make it difficult for the supercomputers to embrace users and ultimately lead to the generally low efficiency of supercomputer applications worldwide.

The application research of the "Yinhe-1" is a visible example of how difficult it is for supercomputers to embrace users. Upon the appraisal of the "Yinhe-1" by the state in 1983, the research on its applications immediately kicked off. When the "Yinhe Petroleum

Seismic Data Processing System" started running in 1987, nearly five years had passed. Five years would have been enough for the world's computers to proceed to a new magnitude of operating speed!

The situation was improved for the "Yinhe-2". It took less than a year between its national appraisal in 1992 and the operation of the Yinhe medium-range numerical weather prediction system in October 1993. An important take-away experience was the establishment of a development team on meteorological application software in parallel with the "Yinhe-2" project, with the purpose of working closely with users and carrying out research on "Yinhe-2" application technology.

Immediately after the launch of the "Yinhe-3" project in 1994, a large-scale application software group was established to conduct research on such critical application areas as meteorological numerical forecasting, computational aerodynamics, nuclear physics engineering and on distributed and scalable parallel algorithms, paving the way for the application of the "Yinhe-3". The team cooperated with users to develop high-level core algorithm software for numerical weather prediction, seismic mechanism research, quantum chemistry research and aerodynamics research, making breakthroughs in the key technology of large-scale parallel scientific algorithms and visualization. The numerical weather prediction spectrum model, computational aerodynamics, MPP and visualized systems with typical hidden forms and distributed parallel processing technology all represented leading technologies in China, reaching the internationally advanced level in the mid-1990s.

Meanwhile, the Yinhe supercomputer practitioners actively looked for ideal partners for the "Yinhe-3", and signed a cooperation intention agreement with the National Aerodynamics Research and Development Center, among other organizations. At the time of the national appraisal of the "Yinhe-3", four organizations approached to discuss service purchases and achieved satisfactory results.

Soon after the appraisal, four "Yinhe-3" computers were produced, representing a fairly impressive performance in the field of supercomputers.

In the following five or six years, the "Yinhe-3" was the most popular and most extensively applied supercomputer in China!

Those involved in the Yinhe supercomputer projects particularly prided themselves on the successful marriage between the "Yinhe-3" and China's military weather forecasting system.

In the age of Cold Weapon, weather forecasting was a critical matter of victory or defeat in the war. During the Three Kingdoms period, Zhuge Liang cleverly used the east wind in the Chibi battle, where a weak and backward force defeated a strong and superior force. The battle became a classic example in the history of human warfare.

Modern warfare, integrating sea, land, air, space and magnetism, requires accurate weather forecasts.

During the Battle of Moscow, the German army launched a surprise attack on the Soviet Union on 22 June 1941. Although the German's long-term meteorological weather forecast had shown that the winter would fall earlier and be colder than previous years in the Soviet Union, Hitler was so obsessed with the "Blitzkrieg" that he ordered a full-scale assault on the Soviet Union without effective cold protection measures and launched the Battle of Moscow in early October. As predicted, the cold wave swept the battle field in early November, earlier than in previous years. A month later, the temperature in Moscow fell to negative 40 degrees Celsius. It was so cold that the German tanks could not start due to the freezing fuel and cooling water and 400–500 non-combat soldiers were off duty in each infantry regiment. In contrast, following the suggestions of the weather forecasting experts, the Red Army Command of the former Soviet Union fully prepared combat equipment and cold protection measures. The Soviet Union eventually secured a victory in defending Moscow, thus reversing the situation of the Second World War.

In October 1996, the meteorological centre of the General Staff Department decided to launch a military numerical weather prediction system project. But who would be the contemporary Zhuge Liang that used the east wind to win the modern high-tech war? At the project demonstration meeting, Li Zezhao, an academic and an expert on meteorology, said, "Go to approach Song Junqiang at the NUDT. He will give you a satisfactory answer."

From December 1996 to October 1997, leaders and staff from the General Staff Department paid multiple visits to the Institute of Computing Technology at the NUDT. They believed that Song Junqiang was indeed the top-notch expert on the application of computer prediction in China and that the "Yinhe-3" was China's leading supercomputer. The medium-range weather prediction system based on the "Yinhe-3" was leading in the world, with its operating time shorter than its foreign counterparts and a speed three or four times faster than its domestic peers.

In December 1997, the Meteorological Center of the General Staff Department and the Institute of Computing Technology at the NUDT signed an agreement on the "Military Weather Prediction

System" technology development and equipment purchase.

It was a long-expected day for Song Junqiang. Early in the development of the "Yinhe-2" medium-range numerical weather prediction project, he led the team to carry out extensive research on military weather prediction and conducted preliminary research to prepare for the development of military weather prediction systems.

He worked together with the team to overcome difficulties and successfully broke through a series of key technology barriers, such as the design and implementation of efficient distributed and parallel algorithms. It took only one and a half years for the team to complete the large-scale distributed and parallel application software system that had cost its foreign counterparts three years. By establishing the first-generation military numerical weather prediction system, it started from scratch and achieved breakthroughs in the medium-range numerical weather prediction of the Chinese army. He later led the team to complete the first phase of the Air Force's aviation medium-range numerical weather prediction system, marking an important milestone in the development of the Air Force's meteorological service.

The achievements were extensively covered by such media as the "People's Daily", "Liberation Army Daily", "Guangming Daily", and "Science and Technology Daily", all speaking highly of the "revolutionary transformation of the military meteorological support means, which significantly improved China's military meteorological support capabilities."

In the summer of 1999, while various meteorological observatories predicted the landing of a strong typhoon on the Chinese mainland, the military weather prediction system of the Meteorological Center of the General Staff Department based on the "Yinhe-3" had made a distinctly different forecast, and through careful calculation determined that the typhoon would not land in China. It later turned out that when the typhoon moved into the waters of the East China Sea a few days later, it suddenly turned north and did not land on the Chinese mainland.

On New Year's Day of 2000, just as people celebrated the arrival of the new century, the long-rumoured "Millennium Bug" computer problem suddenly broke out and affected computer systems around the world. Tens of thousands of computer systems in China were paralyzed. On this day, the "Millennium Bug" visited the National Meteorological Center as feared, resulting in a disorder of certain element storehouses and the failure of weather forecasts for New Year's Day. At this critical juncture, the National Meteorological Center decisively turned on the "Yinhe-3" emergency backup system and successfully resisted the disturbance of the "Millennium Bug", ensuring that the weather forecast was broadcast on time. The chief of the National Meteorological Observatory specially made a call to the School of Computer Science and Technology of the NUDT and extended his gratitude for the development of such a high-level, high-quality supercomputer. The weatherman of the CCTV weather forecaster called it "a special weather forecast."

Since its installation in the Meteorology Center of the General Staff Department, the "Yinhe-3" outstandingly completed the task of weather forecasting for a series of major military activities, such as the naval exercises in the South China Sea, the prediction of a hot air balloon flight orbit and the military parade on the 50th anniversary of the founding of the PRC, and the "997" military operation.

During the 10th Five-Year Plan period, Song Junqiang led the team to make persistent efforts and successively developed the "Global Medium-range Numerical Weather Prediction System" as a key project of the Navy, and the "Boundary Layer High-resolution Numerical Weather Prediction System" as a key project of the PLA Second Artillery Corps. The projects addressed the urgent demand of the Navy and the Second Artillery Corps for meteorological support.

During the 11th Five-Year Plan period, he and the team made painstaking explorations of the global four-dimensional variational data assimilation techniques, which was the bottleneck restricting the development of military weather forecasting technology in China, and solved technical challenges including the multi-resolution incremental four-dimensional variational data assimilation, the direct assimilation of satellite emissivity data and the tangent linear/adjoint modes. The team successfully developed China's first operational four-dimensional variational global meteorological data assimilation system. Aiming at the global forecasting model, the mainstream of international numerical weather prediction studies, they make breakthroughs in a series of key technologies, including the high-resolution global model, and successfully developed China's second-generation global forecast model. Based on this achievement, a second-generation military numerical weather prediction system was successfully developed, with an accuracy rate comparable to that of such meteorological powers as Europe and the United States, creating myths of "predicting the heaven and calculating the earth".

Meteorological conditions are the key to space launch. If good weather is predicted to be bad, the perfect launch opportunities may be missed; and if it is the other way around, the mistake may lead to accidents and disasters.

Before launch of the "Shenzhou 6", the meteorological department predicted three weather events within 72 hours. The forecast of the first weather process was fairly accurate, but the second weather process took place later than the forecast. The satellite image showed that cold air was lingering around the Mazong Mountain area near the launch site and that it tended to converge with the warm and humid air flow down the Tibetan Plateau at any time. While the meteorological consultation was intensified, a strong wind came earlier than expected. According to meteorological personnel, the strong wind would disappear in the early morning two days later. However, no sign emerged that the strong wind would die away. Instead, the wind increased, with a maximum speed reaching 17 m/s, which far surpassed the required wind power for a launch. But the launch would enter an eight-hour countdown in two hours. Would the wind stop then? Everyone could not help but become highly concerned.

The weather forecasters insisted that the 24-hour weather consultation results would remain true and unchanged, believing that the wind would definitely stop on the eve of the launch.

Unexpectedly, after the launch entered the eight-hour countdown process, instead of ceasing, the wind blew with stronger force.

The launch commander asked the weather forecaster in a nearly insensitive and forceful way, "Can you say when the wind will stop?"

Surprisingly, the weather forecaster answered with certainty: "I firmly believe that the weather prediction system is reliable. I guarantee that the winds will definitely stop in a few hours!"

Accordingly, the launch command decided to launch the "Shenzhou 6" spaceflight as scheduled.

At 4:00 a.m., the weather prediction system made a new forecast: A small cloud precipitation system was forming over the northern part of the launch site and was estimated to affect the launch site soon. At 5:00, with the strong winds not diminishing, the small cloud of precipitation brought heavy snow to the launch site. It was quite rare to experience snowfall in October in the Gobi Desert. It was later learned that the snowfall brought by the vortex system was the earliest heavy snow that had embraced the launch site, ever since its establishment over 40 years.

On the eve of the launch, the snowfall came to a sudden end as predicted and the strong wind gradually weakened. During the launch window, the wind died down to a speed of 3m/s, much lower than the required wind speed of 10 m/s.

With a roar, the "Long March 2F" rocket with the "Shenzhou 6" spaceflight aboard lifted off and successfully sent astronauts Fei Junlong and Nie Haisheng to space.

Modern defence technology, especially the aerospace experiments, demands precise quantitative forecasts at certain times and fixed points of atmospheric temperature, pressure, humidity, wind speed, visibility, wind shear and downburst turbulence, among other factors on a global scale. To meet this demand, Song Junqiang led the team to target the exploration of near space. By serving as the head of a number of research projects on the environmental prediction technology of adjacent space, he successfully expanded the scope of military meteorological support from the troposphere to the lower part of near space.

The prediction products of these systems have become the basis of the daily weather consultations and forecast guarantees at all levels of China's military meteorological support departments and have played a critical role in the meteorological support for daily training of troops, major military operations and national defence scientific experiments, among other missions. It played a particularly important role in the meteorological support for major military tasks, such as the rendezvous and docking of the "Tiangong 1" space lab with the manned "Shenzhou 9" and "Shenzhou 10" spacecraft and aircraft carrier trials in the sea.

Chapter VI
Planning Ahead

The ultra-massive parallel processing technology faces many challenges. What is the future of supercomputing technology?

At the crossroads of supercomputing technology development, some are watching, some are wondering and waiting, and others are devoting themselves to exploration.

Yang Xuejun, the chief designer of the Yinhe supercomputer, planned ahead by leading his team to conduct prior research on the supercomputer system architecture technology and coming to the theoretical conclusion of heterogeneous parallel system architecture. The research results pointed the way forward for the future development of supercomputer technology and opened up new and broader space for exploration.

I. The Era of Parallel Computing

The development of computer technology is like a huge river running through the tunnel of time. The major innovations of the overall technology emerge like the spring breeze, bringing about the thawing of the stream in spring and rushing torrents in summer, which are followed by a bleak autumn and silent winter. All technological innovations, including the vacuum tube computers, transistor computers, integrated circuit computers, multiprocessing computers and parallel computers, have gone through such a "seasonal cycle".

In the early 1990s, when multiprocessing technology failed the large-scale scientific and engineering calculations and large-scale data processing, and the world was working towards the massively parallel processing (MPP), the microprocessor (or CPU) was born, leading to a breakthrough in the overall technology of MPP. This breakthrough appeared with an astonishing roar, like the sound of cracking ice that broke the silence of winter and thawed the icy river, and created a brilliant period in the history of computing.

In the course of more than a decade following the overall technological breakthrough of MPP, world supercomputer technology has flourished, witnessing technological leaps and bounds by the United States, Japan and other traditional supercomputer powerhouses and the booming of supercomputers in such Western European countries as Germany, Italy, Britain, France, Canada, Australia, Brazil and India, among other countries. The international supercomputer club has expanded dramatically, producing a number of superstar supercomputers, including:

The "NUMA Cray-T3E" series by Cray;

The "SP2" family of distributed storage architecture by IBM;

The original series of "SGI CC-NUMA" architecture;

The "NAS2000" by NASA;

The "ASCI" by the United States Department of Energy;

The "Earth Simulator" by Yokohama, Japan;

The "NEC" used by the Japanese government;

The "IBM" by the Polytechnic University in Barcelona;

The "Jaguar" by the United States;

The "SuperMUC" by the University of Bavaria;

The "Blue Mountains" by the United States.

It is fair to say that the industry is shining with colourful and dazzling starlight.

With new members rapidly joining the supercomputing club, and with improvements in performance, rapid application expansions, successful applications in strategic simulation and extensive uses in the military, computers have entered the era of supercomputers, and their value has soared.

International strategists believe that "supercomputers have become a strategic area for international competition."

Scientific theories and scientific experiments have been hailed as the two feet of science and technology, which have supported the development and progress of humanity's scientific endeavours. By the end of the 20th century, the booming supercomputers had played an increasingly important role in the advancement of science and technology. Gradually, high-performance computing had come to stand shoulder-to-shoulder with scientific theories and scientific experiments, as the "three pillars buttressing the huge building of modern technology".

Upon this background, Professor Hans Meuer and Eric Strohmaier from the University of Mannheim, Germany initiated the TOP500 project to rank supercomputers globally in 1993. Despite being a purely non-governmental activity, the project was so charming that it attracted attention from around the world, even the favour of politicians. It has gradually evolved into the Olympics for the science and technology sector and has even become a significant platform for the world's powers to show off their computing muscles.

There is no wonder that this period has been called an era of parallel processing in science and technology.

In China, the breakthrough of the MPP overall technology resembled a spring thunder, signalling the advent of the spring of supercomputers.

At the College of Computers at the NUDT, since the MPP breakthrough by the "Yinhe-3", the Yinhe series of supercomputers has continued to make multiple key technological breakthroughs. With blossoming domestic supercomputer brands, the Chinese supercomputer industry has continued to thrive.

To achieve the goals of intelligent computer development under the national high-tech project, the National Research Center for Intelligent Computer Systems was approved by the State Scientific and Technological Commission and established in December 1989. After completing the development of the "Sugon I" and the "Sugon 1000", it successfully developed the "Sugon 2000 I", "Sugon 2000 II", "Sugon 3000", "Sugon 4000L", "Sugon 4000A", "Sugon 5000A", "Sugon Nebula" and other series of high-performance computer systems.

In 1996, to strengthen the research and development of high-end parallel computer systems in China, the National Research Center of Parallel Computer Engineering and Technology was established. In 1999, the Center launched the "Shenwei-1" with an operating speed of 384 gigaflops, and later launched the "Shenwei-2" with an operating speed of 18 teraflops in 2007. It launched the "Shenwei-3" in 2010, which was the first computer equipped with domestic CPUs with an operating speed of 1.1 petaflops.

At the beginning of the 21st century, business-centred Lenovo, Inspur and other companies joined the supercomputer industry and introduced the "Shenteng" series of supercomputers. The "Shenzhen 1800" system launched in 2002, with a measured performance exceeding one teraflops, ranked 24th among the world Top500 in terms of actual computing speed. Launched in 2003, the "Shenteng 6800" ranked 14th among the international TOP500 with an operating speed of 4.183 teraflops as tested by the United States Department of Energy's Berkeley National Laboratory. The "Shenteng 7000", launched in 2007 with an operating speed of 106.5 teraflops and powerful marketing, quickly found its application in education, government, oceanic research and other fields.

Inspur, another major high-performance computer manufacturer, has also successfully developed the "Tiansuo" series of supercomputers and successfully applied it in weather forecasting, petroleum exploration, biopharmaceuticals and other fields.

High-performance computing applications also arose as the supercomputer technology continued to improve. Around 2000, the Shanghai Supercomputing Center and the Supercomputing Center of the Chinese Academy of Sciences were established. In early 2001, the supercomputing centre was completed in the Shanghai High-tech Park and opened to the public, marking a new stage in the development and application of China's high-performance computing.

China's high-performance computing finally embraced the preliminary industrialization!

However, comparing the intergenerational transitions of China's supercomputers and those of the rest of the world, it is clear that the performance improvement of Chinese supercomputers was always followed by the latest and fastest supercomputer developed by the United States or Japan. As noted by the professionals, "China's supercomputers are always a little bit behind those of other countries."

Although it is only a little bit, in the ever-fierce international competition, this can mean a world of difference.

A small advantage can place a nation at the head of the pack; a small gap in development places them in the category "advanced in the world", which means behind the others!

With a small advantage, a nation can stand on top of the peak of achievement to look down upon the whole world, while others can only look up from the hillside.

For China's supercomputer technology to stand out, it is necessary to fight a hard fight.

II. Decisive Occasion

As the gates of the 21st century slowly opened to us human beings, the "parallel processing era" started to encounter a chilling effect. The striking sign of this chilling effect is that the improvement of single-chip performance is greatly slowed under the limitations of the manufacturing process. In other words, scientists can only depend on increasing the scale of the system in order to enhance the overall performance of supercomputers. In this context, there will be a series of insurmountable barriers after the system operation speed achieves petaflops. For instance, the volume will be as large as several football fields, and the requirements of power consumption can be satisfied by building a dedicated power station.

Let us take Japan's "Earth Simulator", which was launched by the NEC Company in June 2004, as an example. Although it had a peak performance of 35.86 teraflops and once ranked first among the international TOP500, it adopted 5,120 customizing multiprocessors and its power consumption reached up to 12 MW. The computer room consists of four floors, such that the machine is stored on the fourth floor, the third floor is equipped with 100 kilometres of copper cable for global interconnection, the second floor is equipped with air conditioners and the power room lies on the first floor. Such a layout is necessary because the power consumption is too high. Although it has undergone changes in programmability and system utility efficiency, the "Earth Simulator" has turned into the opposite of a high-performance computer owing to its high power consumption needs and hardware costs.

With the rapid expansion of volume and power consumption, various problems have appeared, such as difficulties in parallel algorithms design, insufficient communication storage bandwidth, increasing costs in operation and maintenance, inferior system reliability and low safety performance. All these hurdles are unconquerable technological bottlenecks.

The supercomputing technology requires new architectural

theory to support its span. Pure CPU-large-scale parallel computing technology line has thus begun to enter a cooling period.

This means that in the face of new technological peaks in high-performance computers, the progress of supercomputing in developing countries such as China is now at the same starting line as that in the United States, Japan, and other developed countries. Opportunities in which China can win decisive battles in the field of supercomputing and reach the summit of Qomolangma have already appeared!

What is the driving force that will break the ice for future supercomputing technology?

Where is the road to supercomputing development?

Over the course of the history of human computer development, why has the United States been a pioneer, whether in the vacuum tube computer era, transistor computer era, IC computer era, multiprocessing era, parallel processing era, or the massively parallel processing era?

A closer look will reveal that the United States not only succeeded in developing the world's first computer, but has also monopolized the major basic or theoretical innovations of the computer for more than 60 years. Computer components such as vacuum tubes, transistors, integrated circuits, and chips, as well as overall computer structure theories such as multiprocessing, parallel processing, and massively parallel processing, and in particular the three formulas that have given people three leaps in understanding parallel processing—the computational performance speedup formula, the Gustafson's law, and the computer performance model framework—all have been inventions of American scientists. The impetus created by these initiatives, coupled with the increasing traction in computer application demands, have initiated a number of new eras for American computer technology. It has also pushed the United States to become the world's pioneer of computer development and has kept the United States at the forefront of world technology.

The initiation of science and technology, especially the creation of major basic technology and rationales, is a powerful engine for technological and economic development. In the pursuit of technological advancement, the United Kingdom began the modern industrial revolution in the 18th century, the United States accomplished a strong economic rise in the early 19th century, and Germany and Japan quickly recovered from the defeat after the war and regained their status as global economic powers.

Since the founding of New China, and institution of the "reform and opening-up" policy in particular, the nation has attached great importance to scientific undertakings and technological advancement is progressing with each passing day. However, its impetus mostly comes from the introduction of existing technologies, or their subsequent innovations. The authentic technology initiatives, especially the major basic technology and basic theory initiatives that lead the world's scientific frontier and support the nation's industrial transformation, are still rare compared with those of the United States, Japan and European developed countries.

The advancement of Chinese supercomputer technology has always been characterized by "tracking" and "following". The reason why China's technological advancement is always a little bit slower than others is that the initiatives are insufficient.

Only by breaking through the original bottleneck can we truly accomplish the transformation from manufacturing to creating.

In order to transform from "tracking" and "following" to "transcendence" and "leadership" in the field of supercomputing technology, China must establish new approaches to basic technologies and basic theories and chart new paths in as yet undiscovered territory.

When the development of supercomputing technology in the world encounters hurdles, Chinese computer scientists have the responsibility and obligation to make due contributions to the nation's and even to the world's scientific and technological progress!

III. China's Inception and World Initiation

What kind of architecture can overcome the obstacles (high power consumption, large volume and difficult technology implementation) that massively parallel processing supercomputers face?

After some painstaking thoughts and repeated argumentation, Yang Xuejun first proposed the technology of heterogeneous parallel systems all over the world.

The so-called heterogeneous parallel system includes two different types of processors in the computing nodes. One is a traditional general-purpose processor (CPU), which is applied to handle routine tasks, while the other is a customized special-purpose processor that is used to process a specific algorithm. The latter is specially designed to perform well in disposing certain algorithms, thus improving the overall performance of the computing nodes.

But what kind of processors could be used as a special-purpose processor to complete the task of a specific algorithm?

A stream processor named "Imagine", put forward by Bill Dally, Director of the Department of Computer Science at Stanford University, came into Yang Xuejun's sight. Because of his profound academic background and years of practical experience in leading missions, he was incisively aware that this stream processor "Imagine", with many innovative ideas such as the separation of calculations and memory access and development locality was a promising architecture. He intended to apply it with a CPU to supercomputers.

However, the stream processor "Imagine" is merely a research prototype chip. It is generally used only for processing tasks related to streaming media. It is still a mystery whether it can be used to handle scientific and engineering calculations.

To find out this answer, in the year 2006, Yang Xuejun led a stream processor group and hardware and software design team composed of his own students to tackle the problem of scientific computing stream processing technology.

There were three key technical challenges separating stream processors from scientific computing. How could they design the first 64-bit stream processor for scientific computing in the world? How could they rewrite or readjust the applications for this stream processor? How could they efficiently map these applications to the processor?

These obstacles dogged the thoughts of Yang Xuejun and his team members.

No matter how busy the job was, Yang Xuejun would spend two days a week discussing academic issues with the teammates; thus, they often missed meal times. He then generously entertained everyone to a fine meal and continued to discuss topics at the dinner table, often with unexpected results.

Routinely, Yang Xuejun and the team members would carry two mobile phone batteries. Once there were new discoveries, they would call to exchange ideas. Usually the phone call lasted for one or two hours, ending only when both batteries were drained and their ears were burnt.

They successfully broke through a series of technological barriers such as architecture design, program streaming theory construction, and heterogeneous programming model design. They verified the feasibility of stream processors for high-performance computing and proposed the 64-bit stream processor FT64 for scientific and engineering calculations, which was successfully applied to the construction of massively parallel processing systems.

These research results are truly world-firsts.

In June 2007, the stream processor research paper "A 64-bit stream processor architecture for scientific applications" completed by Yang Xuejun and his team was published in the International Symposium on Computer Architecture (ISCA) and was recorded by the international authoritative journal *IEEE Transactions on Parallel and Distributed Systems.* The paper introduced the 64-bit stream processor and its programming methods for scientific computing independently designed by the NUDT. When the IEEE TPDS journal reprinted the paper, the team expanded the dependency-based streaming theory, stream compile optimization methods, and extended experimental data and results.

This was the first academic paper admitted by ISCA from Chinese research institutes and independently accomplished by Chinese scholars. It was also the first architectural theory proposed by Chinese researchers in the history of computer development. The publication of the paper caused a sensation in the computing field at home and abroad.

William Dally, the pioneer of stream processor technology and Director of the Department of Computer Science at Stanford University in the United States, said: "This paper has made significant progress in the development of stream processors in terms of hardware design and programming methods for scientific computing."

Bill Dally, member of the American Academy of Arts and Sciences, member of the National Academy of Engineering, NVIDIA Chief Scientist and former Director of the Department of Computer Science in the Stanford University, praised the research, saying, "This paper has introduced the world's first stream processor used for scientific computing."

Researchers at the University of Wisconsin and the University of Texas published an article at the symposium on microarchitecture "MICRO 2008", saying that Yang Xuejun's research paper on stream processors "described an extensible stream processor for scientific computing applications".

The heterogeneous parallel system of CPU and 64-bit stream processors has provided a brand-new way for the world supercomputing technology to break through the chilling effect.

Chapter VII
Time for a Decisive Battle

In 2008, America adopted the technology of heterogeneous parallel systems and successfully developed the "Roadrunner", the world's first petaflops-capable supercomputer, leading to a new round of decisive battles in the field of international supercomputer technologies.

At the beginning of the 21st century, the Central Committee of the Communist Party of China called for "building an innovative country". The great rejuvenation of the Chinese nation needed powerful support from supercomputers. Innovation in main supporting technologies for supercomputers, such as basic software, network engineering, and microprocessors, boomed with vigour in China.

It was high time for China to wage a decisive battle against international powers in the field of supercomputers.

I. The "Roadrunner" Fired the First Shot in a Decisive Battle

On 18 June 2008, one year after Yang Xuejun published "A 64-bit stream processor architecture for scientific applications", America announced that IBM had adopted the technology of heterogeneous parallel systems and successfully developed a computer named the "Roadrunner", which had a peak performance of 1.37578 petaflops and a LINPACK performance of 1.026 petaflops.

The "Roadrunner" consisted of 6,480 AMD Opteron processors and 12,960 IBM cell processors. The latter were high-performance special-purpose processors. The "Roadrunner" has fully exemplified the outstanding performance of the heterogeneous parallel technology. It has not only significantly improved the performance of each single computing node, but has also dramatically reduced the energy consumption, thus drastically reducing the size of the whole system.

For example, the "Blue Gene/L" supercomputer at the Lawrence Livermore National Laboratory in America, which was among the top 20 systems on the TOP500 list along with the "Roadrunner", had 65,536 nodes. The "Blue Gene/P" system, another supercomputer at IBM, had 73,728 nodes. However, the "Roadrunner" had only 3,240 nodes, which was less than 1/20 of these other two systems. This was due to the cell accelerator, with which the single-node performance of the "Roadrunner" reached 425 gigaflops, while the "Blue Gene/L" and "Blue Gene/P" only achieved 7.3 and 13 gigaflops, respectively. Thanks to a sharp reduction in the scale of nodes, technological bottlenecks in terms of communications, storage, programming, and power consumption were instantly loosened.

The prominent technological advantage of the "Roadrunner" stimulated an upsurge in academic and industrial research on heterogeneous parallel computing.

By taking the lead in the research on heterogeneous parallel systems, the "Roadrunner" fired the first shot in a new round of warfare against world powers in the field of supercomputers.

In the 21st century, China had no choice but to prepare for battle.

Since the institution of its "reform and opening-up" policy in 1978, China's economy has developed at a high speed. In the early 21st century, it successfully replaced Japan as the second largest economy in the world and kept its high speed to gain on America, the world's economic hegemon. At the same time, the modernization and informatization of its national defence and the military had advanced by leaps and bounds. The desire of the Chinese people to realize the rejuvenation of the Chinese nation, as it were, had become more urgent than ever before, and the opportunity had grown riper than ever before.

The country's giant leap from a major manufacturer to a major innovator, the lofty mission of strengthening the military with science and technology, and the great dream of the rise of the nation were in dire need of strong support from such a broad and solid modern platform as that of high-performance computers!

In early 2007, upon hearing that the NUDT had developed a generation of supercomputers, Hu Jintao, then General Secretary of the Communist Party of China, wrote down these instructions: "Our comrades shall further enhance the confidence and courage to scale the world heights of science and technology, continuously improve our capability in independent innovation, and work hard to acquire a number of key technologies in several important fields, so as to make new and greater contributions to the construction of an innovative country and a powerful military strengthened with science and technology."

Faced with aggressive actions by world powers, computer scientists in the NUDT represented by Yang Xuejun responded calmly and bravely. After scientific analysis, they came to the conclusion that, through half a century's continuous effort by developers of the Yinhe supercomputers, the College of Computer Science of the NUDT had already scaled many scientific and technological heights, formed its own characteristics, and accumulated significant achievements in the supercomputer industry. It had not only acquired key technologies for heterogeneous parallel systems that would be the mainstream technologies for the next generation of supercomputers, but also other supporting technological conditions for a decisive battle against other powers for technological domination.

II. The First Condition for a Decisive Battle: The "Chinese Core"

In the mid-1990s, with the development of microprocessor chips, the world took a giant leap from multiprocessing to massively parallel processing in the supercomputing industry. During the development of the "Yinhe-3" supercomputer, researchers represented by Lu Xicheng and Yang Xuejun decided to take the lead in the world by adopting massively parallel processing technology. Thus, they had to use high-performance microprocessor chips, instead of the commonly-used medium- and small-scale integrated circuits.

However, at that time, developing microprocessors was beyond the capacity of China.

Hence, China had no choice but to import microprocessors through all kinds of channels. Moreover, such key technologies as the architecture and instruction system of imported microprocessors were completely locked in. The performance of supercomputers could only be improved by advanced design, which led to a multiplied increase in the difficulty of development. Nevertheless, the researchers could do nothing about the situation.

The embarrassment of having to build "Chinese computers with foreign cores" had deeply hurt the Chinese people.

Thus, professor Li Guokuan from the College of Computer Science of the NUDT made up his mind that he would develop the "Chinese core". His determination coincided with that of two other computer experts, Chen Fujie and Li Sikun. The three wrote a joint letter to the Ministry of Electronics in which they proposed the "Mount Taishan Plan", which intended to develop microprocessors with domestic proprietary intellectual property rights.

At that time, both Western countries and insiders in China thought that, due to China's poor basic conditions and low level of craftsmanship, the Chinese people would not be able to develop their own microprocessors. Developing microprocessors would be like lighting a candle for a blind man, and the "Mount Taishan Plan" would be doomed to fail. But the dream of independently developing CPUs had already sprouted.

Soon, the dream the "Chinese core" finally had enough soil to take root.

During the Eighth Five-Year Plan period, a certain institute under the China Electronics Technology Group had been granted the "DSPC25" project. Although this was just a fixed-point DSP with a few transistors, which seemed to be quite easy, the institute barely finished the project after eight years of strenuous work, during which it had undergone tape-out 15 times and changed researchers three times.

In the Ninth Five-Year Plan period, the institute wanted to carry out the floating-point "DSPC30" (a de facto microprocessor) project. Leaders and experts in the whole microelectronics section doubted they could do it, as the institute had had a hard time in developing the "DSPC25", which was much easier than the "DSPC30" in terms of basic frequency, logical structure, as well as number of transistors. As a result, this project was dragged on and on.

Anxious about the project, relevant department leaders at the COSTIND gave the institute advice, "Since you cannot develop the

‘DSPC30’, why not join hands with researchers in universities?”

However, several well-known universities had rejected the institute’s proposal.

In 1997, Li Guokuan learned this information. He instantly visited the institute along with his staff and talked about cooperation. The two sides chimed in easily and signed a project cooperation agreement in May 1998.

According to the agreement, the College of Computer Science of the NUDT was responsible for logic extraction, logic simulation and test code generation. Although this pre-research project was difficult and underfunded, researchers of the Yinhe team took it as a great opportunity for them to get involved in the microelectronics and microprocessor design field. No matter how small the budget, it was a worthwhile project.

The university established a microprocessor research team, which was led by Li Guokuan and consisted of over 10 researchers. Based on their rich experience in the design and testing of supercomputers, the team, working day and night for over a year, finished the logic design and simulation and produced a correct chip logic netlist.

This scientific chip logic netlist was later used to crosscheck the logic and layout design of the “DSPC30”. A large number of mistakes were identified and revised in the layout design. As a result, the project created a miracle of successful tape-out once and for all, and performance of the chip was stable and satisfactory.

The DSPC30 project was rated as an outstanding national defence project and hailed as "a model of joint research between universities and institutes".

The success in one stroke had overwhelmingly encouraged the microprocessor research team to reach for higher goals.

In the early 1990s, China started the research on its third-generation fighters. Researchers were resolved to achieve a series of breakthroughs in key technologies of digital aviation. However, just when they had produced several prototype fighters, America suddenly announced an embargo on the microprocessors that were indispensable to the third-generation fighters. As a result, China's new generation fighter project was instantly stuck.

In order to realize the historical leap of China's military aircraft from the second generation to the third generation as soon as possible, the General Armament Department immediately started a research project on microprocessor chips for fighters, and appointed the College of Computer Science of the NUDT and the State Microprocessor Corporation to design the chip simultaneously.

Each of the two organizations designed one of the two chips of the microprocessor, but the main chip was designed by them both simultaneously. Obviously, this was a player-killing competition. Only the chip with better performance would be adopted.

Although Li Guokuan's team had previously participated in the DSPC30 project, they had only undertaken its logic design and simulation. In this mission, they had to carry out the layout design and simulation, after which the design would be directly used in platemaking, taping out, packaging, testing, trial, examination and finalization of the design. As foreign countries imposed strict blockades on relevant material, domestic researchers had to rely on their own efforts in the development of the system structure, layout design and implementation technique.

The microprocessor research team of the NUDT bravely rose to the challenge. After a long period of scientific and technological research, they had cultivated a kind of psychological inertia—where there was a challenge, there was an opportunity for innovation!

At the kick-off meeting of the project, the researchers declared with their spirits soaring to the firmament, "We must develop the 'Chinese core' for China's new generation of fighters!"

After reviewing the development plan, experts from the General

Armament Department came to the conclusion that this was the most difficult microprocessor development project in China. They suggested dividing this mission into two stages: in the first stage, researchers would develop a complex instruction set superscalar microarchitecture and logic design technique; in the second, they would work on a full custom design of ultra-deep submicron scale.

Such an arrangement was definitely reliable, but it took too much time. And time is the most precious and unaffordable element in the modernization process of the Chinese military.

After much deliberation, the microprocessor research team of the NUDT chose the tactics of coordinated actions by multiple groups, formulating unified norms to guide operations of the whole research team. This would allow them to break down the difficulties and solve them one by one, establish an advance detachment to carry out pre-research on commonly used technologies, undertake advance research, design, analysis, and even tape-out for new technologies, structures, and circuits ad independently develop several automatic tools to speed up the design process and improve the efficiency. Over 100 researchers, from the chief designer to layout designers, worked seven days a week, and over 10 hours each day.

After five years of hard work, this type of chip successfully taped out once for all and set three records in China, as the first Chinese microprocessor with a superscalar CISC architecture, the most complicated and massive Chinese microprocessor with a full custom design of ultra-deep submicron technology and the first full custom-designed Chinese microprocessor with sequential modelling and sequential analysis technology.

In December 2001, China's new generation of fighters, equipped with a "Chinese core" designed by the microprocessor research team of the NUDT, proudly soared into the sky over its motherland. The test flight proved that each tactical and technical indicator of this third-generation fighter was by no means inferior to that of other third-generation fighters in the world.

The curse of having no independently developed military chips for China was miraculously broken!

This achievement won a first prize of the Military Science and Technology Progress Award in 2007 and a second prize of the National Science and Technology Progress Award in 2008.

The successful test flight of the third-generation fighter had ignited the Chinese people's hope for the "Chinese core". The microelectronics subject in the NUDT has also established its reputation. The university successfully set up a master and a doctoral programme for microelectronics in one fell swoop and established its leading place in microelectronics design technology in China.

The microprocessor research team of the NUDT was like a precious seed in spring, which finally bloomed and bore fruit across the microprocessor field in China.

A certain institute under the China Electronics Technology Group had cooperated with Russia in the development of the "386" chip. After two years of research, a lot of money had been spent, but the project had seen no progress at all. A careful inspection found that, surprisingly, Russia wasn't able to develop the "386" chip. The Russians' intention in cooperating with the institute was to make use of its money to develop a different product that they urgently needed.

Therefore, the leaders of the General Armament Department wished the NUDT to provide technological support for this institute, help them develop the "386DX" and "387DX" chips, and lead them into the microprocessor field.

The microprocessor research team of the NUDT proactively lent a helping hand by sending capable researchers to work together with those of the institute, and successfully developed these two chips.

At that very moment, a strategic weapons project in the country was in dire need of a certain type of chip. In the beginning, the organization in charge of this project wanted to import the chips, but they were found to be under an embargo.

Hence, at such a difficult time, the microprocessor research team of the NUDT was once again entrusted with the mission of designing the chips, and a certain institute under the China Electronics Technology Group was in charge of the follow-on production of the chips. Both sides worked closely and succeeded in the development of a "Chinese core" for the country's strategic weapons. The institute also made use of this opportunity to get into the country's microelectronics and microprocessor field.

During the 10th and 11th Five-Year Plan periods, the team had not only provided many domestic institutes with technological support and worked together with them in the development of several types of chips, but had also constantly sought technological innovation.

Chen Shuming, the team's second-generation academic leader in the microelectronics field, often vividly spoke to members of the team, saying, "We shall treat new technology as sensitively as young women pursue fashionable clothes. And if we are to keep this kind of sensitivity, we must stand higher and look further than others!"

In order to follow the fashion of the microelectronics field, he always piled heaps of over 10 different top microelectronics journals on his desk, and stuffed his bookcase with all kinds of papers. Through these, he kept close tabs on the forefront of his field.

In 2004, Chen Shuming found that the country and military were in urgent need of chips intended for the space programme, which were mainly imported from foreign countries. He had foreseen that, sooner or later, foreign countries would embargo these chips, just as they did during the development of the fighter jets.

In order to help China's aerospace industry avoid the same mistakes committed in the development of chips for use in the fighters, Chen Shuming instantly lead his team to carry out research on the irradiation effect of integrated circuits, a key technology for chips for use in space, and made a series of original achievements.

As expected, foreign countries soon blocked all the channels through which China imported chips for use in this kind of endeavour.

Then, with the support of the Core Electronic Devices, High-end Generic Chips and Basic Software Project, a project pursuing the development of the first domestic high-performance anti-irradiation digital signal processor was approved. As the person in charge of this project, Chen Shuming started to lead his team develop the first domestic high-performance DSP chip for use in space. With three years of arduous work, they successfully developed a space-worthy "Chinese core" in time, safeguarding the further development of a series of aerospace projects in China.

In recent years, the microprocessor research team of the NUDT independently developed the "586" microprocessor, the "DSP 3000" series, the "DSP 5000" series, and the "DSP 6000" series microprocessors, which were the most advanced in China, and completely broke the foreign blockade on key technologies of Intel's "586", "486" and "386" series of microprocessors and TI's "DSP" series of microprocessors.

Finally, a flourishing spring for the independent innovation of Chinese microprocessors had arrived.

As a main force in the development of the "Chinese core", the microprocessor research team of the NUDT rapidly grew strong during its march toward the forefront of science and technology.

In 2004, the Microelectronics and Microprocessor Institute of the National University of Defense Technology was officially established.

In 2005, the Innovation Center for Microelectronics and Microprocessor Masters was added to the "Education Revitalization Plan" by the Ministry of Education.

In 2006, the technological innovation team for high-performance microprocessors was selected as an innovation team of the Ministry of Education.

This unit, which originally consisted of just over 10 people assisting other institutes in logic design and simulation, has since grown into a strong team with over 100 capable researchers, over 100 individuals with doctorates and master's degrees, and over 100 scientific and technical personnel.

With such strong teams and rich technological accumulation, China has the confidence to declare to the world that the time in which China completely relied on imported microprocessors is gone forever!

III. The Second Condition for a Decisive Battle: "Chinese Kylin"

When it comes to basic computer software, there have been lots of metaphors.

Some say if that if the computer hardware is compared to a human being's body, then basic software is the brain.

Some say if that if the computer hardware is compared to a nation's economic foundations, then basic software is the nation's superstructure.

Perhaps even more vividly, if the computer hardware is compared to an abacus, then basic software is the arithmetic table.

A computer without basic software is indeed a human without

a brain, an economic foundation without a superstructure, and an abacus without arithmetic tables.

Such is the status of basic software in computer systems.

Basic software plays a role in managing, allocating and controlling all computer resources to provide all sorts of application software in the environment of development, allocation and management. In particular, the operating system is the core software of a computer system. Any new technology in terms of computer hardware must be displayed via the basic software, while any advancement in computing applications must be supported by basic software.

For more than 10 years after the birth of the first digital computer, computers were operated without any basic software and were called "naked computers". It was not until the 1960s that basic software such as operating system sand compilers appeared.

The PLA Military Institute of Engineering (the predecessor of the National University of Defense Technology) pioneered basic software research, including operating systems and FORTRAN compilers, across the nation at the end of the 1960s. In the 1970s, it developed various management programs for the "441B-III" general-purpose transistor computer and the "151" mainframe, realizing time-sharing operating and batch processing for multiple users, which was China's first computer operating system. The NUDT played a leading a role in the continuous development of this technology in the basic software field afterwards. It developed and completed the powerful operating system of the "Yinhe-1" and "Yinhe-2" between the end of the 1970s and the beginning of the 1990s. The supercomputers supported by the above systems were rewarded the special science and technology progress award of China's central military commission and the first prize of the National Science and Technology Progress Award. Since then, the operating system research at the College of Computer Science of the NUDT has entered a period of development in line with international standards. In 1993, it cooperated with the British Unisoft Corporation and successfully developed the world's first Unixwane 2.1 operating system that supported Powenpc. In 1997, it successfully developed the world's first massive parallel processing microkernel operating system supporting the "MIPSR4000". The "Yinhe-3" installed with the operating system achieved a major breakthrough in China's supercomputers from one gigaflops to 10 gigaflops, and the basic software innovation team became the leading player in this technical field in China.

Satellite positioning and navigation technology is an important factor of modern informatization warfare. The United States has not only used its Global Positioning System (GPS), which was the first of its kind in the world, to strike out a series of precision attacks during the Iraq War, the Kosovo War and the war in Afghanistan, but also threatened to cut off GPS signals to many developing countries, including China.

Satellite positioning and navigation technology has become a key technology that decides a country's future and a nation's ability to thrive and survive.

With the aim of changing China's passive status in the satellite positioning and navigation technology field and in order to promote the informatization development of the country and the army, China initiated the programme of the "Beidou" satellite positioning system.

The Party Committee of the College of Computer Science believed that the "Beidou" project was a major project in the army's informatization construction and that the College of Computer Science, as the unit involved in the development of the Yinhe series of supercomputers, should actively participate in the construction of the project and make use of profound technical accumulation to make direct contributions to improving the combat effectiveness of China's military.

The "Beidou" information processing system is an important part of the positioning and navigation system. It undertakes the core tasks of data processing, information exchange, system management, and business management of the entire large-scale system. As it is the nerve centre of the three major functions of ensuring system-wide positioning, communication, and timing, people vividly described it as "the brain of Beidou". The Party Committee of the College of Computer Science actively organized its strength to participate in national bids and succeeded in winning the key task of the "Beidou" project. It signed a contract for the development of the information processing subsystem of the "Beidou" ground application system with relevant departments of the General Staff Headquarters.

The College of Computer had established a technical research team with Wang Zhiying as the chief designer, and with Jin Shiyao, Zhao Long and Zhu Haibin as deputy chief designers concurrently serving as principal designers of the three subsystems.

The entire research team spared no effort and went to Beijing, Wuhan, Xi'an, Shanghai and other places for in-depth research and investigation and completed a general research proposal in May 1997. The review experts of the General Staff Headquarters commented that "the overall proposal is technologically advanced, scientifically demonstrated, strong in engineering, and effective in implementing measures to meet the technical requirements for the development of information processing subsystems."

After the programme was kicked off, the entire team worked day and night with a strong sense of responsibility to win the battles for the army and to make contributions without taking any weekends off. They carried forward the spirit of indomitable fighting and rapidly accelerated the engineering progress.

In May 1998, the overall design and interface of the information processing subsystem were completed.

In August 1998, the overall design of the information processing subsystem passed the joint review of the general station of satellite positioning, the General Armaments Department's Measurement, Tracking and Communication bureau, Beijing Aerospace Command and Control Center, Wuhan Institute of Physics and Mathematics, the Chinese Academy of Sciences as well as the Surveying and Mapping Institute of the General Staff Headquarters.

In September 1998, the subsystem of information processing carried out joint debugging of the internal timing, positioning and DEM application programs, all of which achieved good results.

The subsystem of information processing went into the stage of joint debugging in February 1999.

It went into the trial operation stage in September 1999.

...

In 2000, the first "Beidou" experimental satellite was launched towards the sky. This event marked the completion of the "Beidou-1" project and the satellites started to play a role in disaster relief and rescue, transportation, land mapping and military operations.

On 28 December 2011, China announced to the world that the "Beidou" satellite navigation system would provide global navigation services to the world, as the fourth largest global satellite navigation system.

The application proved that the information processing subsystem developed by the College of Computer Science could completely meet the requirement of "fast capturing, monitoring and processing". People applauded them for "developing a smart brain for Beidou".

While China's basic software was pressing ahead, basic research and development in the Western countries was advancing at a rapider pace. The fast pace and leapfrogging upgrades of Microsoft's operating systems were bewildering.

In order to change the game and provide a self-developed

operating system for the national citizens, the national "863 Program" set up an important software project in 2002, which specifically identified the operating system as the major research direction for the primary software project, with a focus on the "Kylin" operating system.

Led by the NUDT, the project of the "Server Operating System Core" was officially kicked off with the collaboration of ChinaSoft International, which has long-term technical accumulation, and other companies such as Lenovo and Inspur, which have a market edge in the promotion of operating system. The project aimed to develop the core of a Chinese server operating system with independent copyrights, high performance, wide application, strong security and compatibility with the Linux binary application to meet the application requirements of the national defence, political affairs and other critical fields.

In April 2005, the subject team rolled out the "Kylin" operating system version 1.0 and version 2.0 with the primary functions of a general server operating system after nearly three years of effort.

In May 2005, the expert panel of the "863 Program" primary software project asked the Weapon and Equipment Certification Center of the General Armaments Department and China Software

Testing Center to accept and test the operating system, concerning 11 projects including instalments, functions, performance, reliability, expanded ability, standard compliance, high practicability, and Chinese language support. After seven months of rigid testing, it was concluded that the "'Kylin' server operating system version 2.0 met all technical indicator requirements of the subject task contract".

Finally, the Chinese people could use a trusted and reliable operating system!

However, faced with such great news, some people wrote an article titled "A Similarity Analysis Between the Kylin OS Core and Other Operating System Cores", the motivation for which was unclear, which raised suspicion of "863 Kylin operating system plagiarism".

To clarify the confusion, the "863" Information Technology Office of the Ministry of Science and Technology established a coding review team led by Gong Min and carried out a 10-day material and coding review of the "Kylin" operating system version 2.0 in August 2006. In December 2006, the "863" Information Technology Office of the Ministry of Science and Technology carried out another acceptance review of the "Server Operating System Core" (Kylin OS) in Beijing.

Both reviews led to one conclusion: all of the technologies of the "Kylin" operating system were indigenous!

At the end of 2008, the subject of a "military server operating system" for the major scientific and technical project under the national "Core Electronic Devices, High-end Generic Chips and Basic Software Project" was officially kicked off, led by the NUDT with the collaboration of the China Standard Software Company. Guided by the principle of "military and civil combination, general civilian and specific military applications", the innovation team developed a customized military server operating system with high-level security and practicability, which effectively bolstered the army's comprehensive electronic information system and realized security control and independent protection of the army's basic software.

Continuously supported by the "863 Program" primary software project and the "Core Electronic Devices, High-end Generic Chips and Basic Software Project", the "Kylin" operating system has developed a series of brands:

The "Yinhe Kylin" series of high-performance computer operating systems, issued by the NUDT, supports heterogeneous parallel architecture, 64-bit multi-core and multithreading microprocessor and SoC architecture, advanced routing high-speed interconnection communication, provides multi-level parallel compiler optimization support and high-performance virtual computing domain management capabilities and software and hardware integrated low-power control technologies which have enabled an integrated energy management framework, making significant contributions to the continuous leap of the national supercomputer.

The "Kylin" series of general-purpose operating systems includes multiple different versions, of which the military version is published by the NUDT, while the civilian version is published by the China Standard Software Company. This series of products independently designed a structured protection-level security mechanism and an entity-integrated mandatory access control mechanism, focusing on system-level fault-tolerant, highly reliable I/O storage, checkpoints, and cluster high-availability support for optimized design, focusing on well-constructed hardware and software environments, taking into account the friendly human-computer interaction interface. The military version of Kylin OS has been successfully applied in military information systems and weapons and equipment.

The "Ubuntu Kylin" series of open source operating system was jointly developed by the NUDT and the Ubuntu community. It adopted a combination of community mode and closed development to break through the unified UI design of the desktop, automatic software package updating technology based on software depot, and customized Chinese language with the integration of a desktop, plug-in system management and framework maintenance technology. The "Ubuntu Kylin" series of operating systems is the first Chinese open source operating system series developed by the Chinese team and officially recognized by the international community. It has been reported on by the BBC, OMG, Xinhua, CSDN and other major domestic and foreign media outlets. It was successfully applied in the China Aid Projects of the Ministry of Commerce and has been exported to over 30 countries and regions to date. Factory installations have been completed in products from Lenovo, HP, Dell and other hardware manufacturers on a small scale. A strategic partnership was established with JINSHAN WPS, which adopted bundling sales and realized wide application in education, electric political affairs and other fields.

Mac, Linux, Windows and Unix rank among the four mainstream operating systems in the world.

Many computer experts who used the "Kylin" operating system observed that "our Kylin OS combines the efficiency of Mac, the openness of Linux, the convenience of Windows and the security of Unix."

The Kylin is a mythical creature known in Chinese legends that has a good temperament and could live for 2,000 years. In the eyes of the Chinese, a place will be deemed as auspicious land if a Kylin appears.

If the "Chinese Kylin" has come, can the Chinese supercomputer boom be far behind?

IV. The Third Condition for a Decisive Battle: "China's First Network"

With the increasing popularity of computers, people are increasingly dissatisfied with working on their own systems, and they are eager to share information and computer resources with computer users in different locations (regardless of distance) and discuss and solve the same problems. These desires can be realized by connecting computers distributed in different places and units.

As a result, the first computer network came out in 1969, and thus opened the prelude to the computer network era. In the 1980s, the Internet came into being, ushering in a period of fast development of computer networks, which gradually developed into one of the fastest, the most influential and most popular technology for mankind.

Early in the 1980s, computer experts of the NUDT acutely realized that the emerging technology would guide human being towards the “digital era” in the near future. They took the lead in setting up a computer network research lab in the country to conduct research on network technology, and obtained research results on the Yinhe front-end machine software and Ethernet technology, which played an important role in the successful application of the Yinhe supercomputers in fields of petroleum, earthquake prediction, and meteorology.

At the same time, the university dispatched two young experts, Lu Xicheng and Dou Wenhua, as the first batch of Chinese network technology students to go to the United States, where the network technology originated. Gong Zhenghu was sent to study in the United Kingdom, majoring in network engineering. In 1984, after returning from their studies, they did a lot of pioneering work and jointly wrote a report to the COSTIND, recommending that they begin the research of military comprehensive computer networks as soon as possible, a project which was highly valued by the competent leaders and departments, and which was provided with funds in the form of an additional project during the Seventh Five-Year Plan period. The university began to comprehensively discuss technologies such as the Yinhe supercomputer network, military network, OSI (open system interconnection) network protocol, and "X.25" communication protocol. The university implemented the OSI protocol suite for the first time in China and mastered the network software implementation technology in mainstream operating systems. In 1988, 1989, 1992 and 1993, the research results won the second prize of the ministerial-level science and technology progress award in a row.

When the "Yinhe-2" project was established, there was no network technical requirement for development in the contract. However, Lu Xicheng and Dou Wenhua requested the addition of a high-speed computer network to the machine and received strong support from the project leader, Chen Fujie, and the chief designer, Zhou Xingming. In combination with the Yinhe supercomputer project, they boldly attempted innovations in the development of a high-speed computer network. For the first time in China, they enabled supercomputers to use a high-speed fibre network via FDDI (Fibre Distributed Data Interface) and virtual station software for networking. This technology was awarded the first prize of the Scientific and Technological Progress Awards of COSTIND in 1993.

In 1993, the network research group developed into a computer network and communication research lab, and the network technology innovation had entered the "fast track."

The Yinhe computer network project successfully developed the first domestic computer network capable of supporting a supercomputer in 1997, which enabled the Yinhe supercomputers to have direct high-speed networking capabilities. This achievement won the first prize of the military science and technology progress awards. The research lab also shouldered the responsibility for the construction of the NUDT campus network (a civilian network) and military area network, and completed two phases of construction in 2000 known as "China's first campus network". Accepted by the Ministry of Education, the networking technology had reached the domestic advanced level.

From 1993 to 2000, the research lab received a total of two first prizes and four second prizes of the ministerial-level science and technology progress awards.

In 2000, the College of Computer Science established the Institute of Network and Information Security, while innovation in the discipline had entered a stage of rapid development.

First of all, the research staff continued to focus on the projects of high-performance computer systems and carried out in-depth development of supporting computer networks.

Right after the Yinhe network system was established, the innovation team successfully developed a high-performance network system that could support a speed at teraflops in 2000. It achieved breakthroughs and innovations in high-capacity network information exchange platforms and system single-image technologies, which achieved the first prize of the military science and technology progress awards.

In 2007, they successfully developed a high-performance network environment for the parallel processing system of Yinhe supercomputers, realizing the high-speed interconnection of the supercomputer's internal network and the TCP/IP computer network.

In 2008, they began to independently develop a high-performance IB (InfiniBand) switch for the future network requirements of supercomputers capable of 10 petaflops. They had prepared high-level network technologies for the new generation supercomputer peaks and were ready for decisive battles with the computer powers of the world.

Next, they launched a fierce rush towards developing the core technologies of computer networks.

In 1999, the country began to implement the "Strategic Plan for the Development of the Internet". With the support of the "211" project, the "985" project and the national "863 Program" key projects and the army's Ninth Five-Year Plan, 10th Five-Year Plan and 11th Five-Year Plan, the network technology innovation team seized this major opportunity to combine independent innovation and integrated innovation, and had formed the scientific research layout for the coordinated development of basic research, preliminary research and model engineering and had conquered a number of sequential challenges in network technology.

A core router is the core component of a network system. If the network system is like complicated, connected and densely packed blood vessels, then the core router is the human heart.

In March 2001, the team successfully developed China's first core router, the "Yinhe Yuheng 9108". The appraisal committee, with the academic Wu Hequan, Vice President of CAE, at its head believed that "'Yinhe Yuheng 9108' is the first high-end wirespeed core router in China that has both self-developed hardware and software. The overall technology is leading in China and reaches the international advanced level. It is another major achievement in China's high-tech field".

In March 2004, the team successfully developed China's new generation of high-performance Internet router, the IPv6 router. The appraisal committee unanimously observed that "this is the first domestic IPv6 router with an exchange capability of over 100 billion bits per second. It has independent intellectual property rights, has achieved breakthroughs and innovations in the development of router parallel architecture, global flow control mechanism, IPv6 wirespeed forwarding, and multiple levels of service quality control, PKI-based network security certification and other key technologies. The overall technology is leading in China and has reached the international advanced level, marking a new level in China's development of IPv6 high-performance network equipment and playing an important role in accelerating China's high-speed network building as well as enhancing security of China's Internet information". This project won the second prize of the National Science and Technology Progress Awards in 2006.

After 2007, they successively launched China's first 2.5G, 10G and 40G network security supervision series with different levels of needs, which were widely used by the state and the relevant departments of the military to provide technical means and play important roles in the crackdown on cybercrime, stability maintenance and counter-terrorism, as well as in the clean-up of Internet space.

Their key research subject of the "Next Generation Network Architecture Model and Research on Super High-Speed Network Switching Routing" supported by the National Natural Science Foundation is a landmark achievement in the development of China's network technology. The acceptance expert group led by the academic Fang Binxing said that "the subject has made major innovation achievement and remarkable progress in the next-generation Internet architecture, high-speed network switching methods and algorithms, large-capacity routing table organization and management, network processor architecture technology, and 10Gbps high-speed network interface linearity as well as its implementation."

This series of network routers had won the second prize of the National Science and Technology Progress Awards several times.

Again, they carried out a series of cyber security technology research projects based on the urgent needs of national and military cyber security.

In the 21st century, network security issues such as computer viruses, network stealing, and denial of service attacks had become the biggest challenges looming over the development of the Internet.

National and military strategic information urgently required security protection and supervision equipment. The development team began to assume various types of network security equipment model tasks in 2005, and successfully passed the design and finalization in 2007. Their products were used by the General Staff Department, Nanjing Military Region, Guangzhou Military Region, the Navy, the Air Force, and the Second Artillery Corps. They were proved to be reliable and safe during a number of drills and Internet combat field exercises. In response to the urgent needs of the national security department, they successively developed a variety of network security monitoring equipment with a trunk line speed to ensure that the security sector could make a fundamental shift from incompetence to general control of online crimes.

This series of equipment and technological achievements won the second prize of the National Science and Technology Progress Awards in 2008.

In addition, they also focused on the sustainable development of discipline innovation and achieved a series of basic network research results.

In 2003, they undertook the "new generation of Internet routing and switching theory" subject research within the frame of the "973 project", and made important progress in the development of cluster router architecture, the packet exchange model, and inter-domain routing stability research. In 2008, the research project was rated as "excellent" in the acceptance inspection organized by the Ministry of Science and Tcchnology and obtained the continuous support of the "973 project".

In March 2009, in collaboration with multiple units inside and outside the armed forces, the network institute undertook a major basic research project on national security, and achieved a breakthrough in terms of the "973" project for the College of Computer Science. The project focused on three scientific issues, which were security invulnerability measures for military networks, the aggregation effect and evolution mechanism of military cyber threat behaviours, and the reconfigurability of military networks and adaptive methods of network components. The project made major breakthroughs in basic theories through innovative research.

These basic research achievements not only provided solid support for follow-up scientific research projects and model projects of high-performance computer networks, high-end network equipment, strategic information systems security equipment, space-based information systems, and data centre networks, but also published over 300 papers registered in Science Citation Index and top international conferences and academic journals such as Sigcomm and Infocom, which greatly increased the influence and popularity of the scientific research team at home and abroad.

Through 30 years of arduous innovation, the security routers, network security protection technology and network security supervision technology they developed had met the requirements of all or different levels. More than 1,200 sets of equipment had been widely used in information systems such as those of the armed forces, the Ministry of Public Security, the Ministry of Security and the Ministry of Industry and Information Technology, to better solve the problem of security interconnection between strategic networks and command and control platforms, weapon platforms and sensing platforms, making contributions to the national safety supervision.

After a series of major scientific and technological breakthroughs, the cause had brought together talents and formed a network communication innovation team centred on scientific and technological elites such as Lu Xicheng, Wang Huaimin, Su Jinshu, Lu Zexin, Wang Baosheng, Sun Zhigang, Chen Shuhui, Xu Ming and Wang Yongjun. The network technology team led by Su Jinshu had been selected for the "Hunan Provincial University

Science and Technology Innovation Team" and the "Innovation Team Development Plan" of the Ministry of Education, thus becoming a national team of network technology innovation. The network technology laboratory led by the academic Lu Xicheng was identified as a key laboratory of the military through the evaluation of the General Staff Headquarters. The laboratory had the most advanced network communication equipment in China, and could build various network demonstration environments and experimental platforms. It also undertook key tasks such as the development of military network interconnection equipment, security protection equipment and the research and application of systems.

These demonstrated that the NUDT had become a national base for computer network research, application development and dissemination, a base for gathering and cultivating talents in this field and a base for promoting the development of computer networks and related industries.

Their profound technical reserves and hard-working strength have won the network technology innovation team of the NUDT the reputation of "China's First Network".

V. The Fourth Condition of the Decisive Battle: "China's Number One Supercomputer Team"

The growth of scientific and technological talents requires the exercise of scientific research and the nourishment of scientific and technological knowledge. Teaching classes and undertaking scientific research are the "two wings" of scientific and technological talents. Only when the "two wings" rely on each other and work together can the scientific and technological talents fly like an eagle in the broad sky of scientific and technological innovation.

In the late 1950s, the PLA Military Institute of Engineering began China's computer science education by establishing the computing discipline. After that, during its more than half a century of teaching work, it persisted in the development path of "relying on scientific research, facing the world and advancing with the trend of the times" and formed a positive cycle in which "teaching and scientific research were mutually supportive and achieved sustainable development together". And it has created the glory of "achieving four times of leap and training five generations of elites".

In 1958, the "901" computer that was designed as a warship control centre was developed successfully at the PLA Military Institute of Engineering. After a few months, following the suggestion of Ci Yungui, Chen Geng established a computer teaching team in the Department of Naval Engineering of the PLA Military Institute of Engineering, led by Hu Shouren. Soon after, the teaching team developed into the electronic computer teaching and research office which achieved a leap in China's computer science education, and which was starting from scratch.

In August 1961, the PLA Military Institute of Engineering established the Department of Electronic Engineering and a military-use computer teaching and research office section was created to mainly serve the army and air force. Meanwhile, the "901" torpedo speedboat director development team also expanded into the navy missile director system teaching and research office. The two teams shouldered the professional teaching work of the college's computer science and technology major and initially shaped a quite complete teaching system. In June 1961, with the authorization of COSTIND, the computer science and technology major in the PLA Military Institute of Engineering started to enrol graduate students. As the first graduate supervisor, Ci Yungui started to recruit graduates from the computer science and technology major that year. The computer science and technology teaching department in PLA Military Institute of Engineering entered a rapid development period and within five years, it had enrolled 368 undergraduate students and increased the number of teachers from nine to more than 80.

In the mid-1960s, with the appearance of China's first transistor computer, the "441-B", the teaching work of the PLA Military Institute of Engineering had their first landmark achievement, which had two important aspects.

The first of these came in the form of a research and development team, with Ci Yungui as its core and elites such as Hu Shouren, Liu Kejun and Kang Peng as the backbone of the team.

Second, the college established the first Department of Computer Science and Technology among the national universities in April 1966. General Nie Rongzhen decreed the order in person and appointed Ci Yungui as the head of the department.

Regrettably, when the computer teaching work sailed forward rapidly, it hit rough waters almost immediately.

Fortunately, because of the urgent military needs plus the natural relationship between the college and the defence institutions, the college's computer department still shouldered heavy work of developing special-purpose military computers. Many undergraduates and junior college students started to focus on the development of certain prototypes as soon as they graduated. Some even were pushed to participate in the research and development team before they graduated. Through the hard honing of scientific and research practices, some major milestones were achieved, such as China's first megaflops-capable computer developed for the "Yuanwang" oceangoing vessel and the first supercomputer, the "Yinhe-1", and a group of young research elites emerged quickly. They contributed creatively in the development of the "Yinhe" series of supercomputers, national microelectronics, basic software, internet technology, computer applications and intellectual technology.

In 1977, China restored the college entrance exam system and the education industry started to bring order out of chaos. In 1978, acting on the concern of Deng Xiaoping, the college returned to the army and was reconstructed as part of the National University of Defense Technology. Under the guidance of Qian Xuesen, the Deputy Director of COSTIND, the university established departments for the various disciplines, integrating science and engineering with engineering as the priority. It was a new start as the National University of Defense Technology. To improve the quality of teaching and training, the teaching work of the university had to be "standardized".

According to the requirement of "setting departments by disciplines", the Department of Computer Science and Technology established the two teaching and research offices of computer applications and computer software. Although the number of teachers on the teaching team was small, everyone was an expert with long-term teaching and scientific research experience or had transferred from famous universities like Zhejiang University, Fudan University, Tsinghua University, Peking University, Nankai University, Northwestern Polytechnical University, Jilin University, University of Science and Technology of China or Wuhan University. These teachers had strong personal abilities, professionalism and team spirit. However, because they had not engaged in teaching for many years, they generally lacked teaching experience.

One night, the teachers were working overtime in the teaching and research office. Ci Yungui came to see everyone and said earnestly, "We have done a good job in scientific research. Everyone says that we are the 'Japan' in the 'four little dragons' of the Chinese computer industry (Japan, South Korea, Singapore and China Taiwan were known as 'Asian Four Little Dragons' due to their rapid economic development. The 'Four Little Dragons' in the Chinese Computer Industry were the four first-class computer scientific research institutions, including the Institute of Computing Technology of the NUDT, the 15th Institution of Ministry of Electronics, the Institution of Computing Technology of the Chinese Academy of Sciences and the East China Institution of Computing Technology). However, we are weak in teaching and lack fame both inside and outside the university. Thus, the Party Committee decided to deploy a number of key personnel to strengthen teaching and strive to make teaching as good as the scientific research."

To make the teaching as good as the scientific research, the teaching team innovated and bravely forged ahead.

In the 1950s and 1960s, China's computer education was influenced by the former Soviet Union. The teaching content mainly consisted of digital electronics, analogue electronics, digital logic, computer principles, computer architecture, program design and calculation methods. After the institution of the "reform and opening-up" policy in 1978, they found that the computer education professional curriculum system in Western developed countries, represented by the United States, had undergone great changes, and many new courses had been created, such as classes on data structures, discrete mathematics and databases. Some of the courses were on subjects they had never heard of, and although some course subjects were familiar, the content was unfamiliar, such as the principle of the compiler and operating system.

The teaching team faced the difficulties. Through hard research and preparing lessons for the students and by preparing a series of new courses, it was initially in line with the international standards. Meanwhile, they insisted on the combination of teaching and scientific research by participating actively in "Yinhe-1" and moving the results back to the platform in time to train talents with advanced discipline knowledge. As a group of 1977 and 1978 graduates participated in the teaching positions, the team basically consisted of old, middle-age and young teachers and became more dynamic.

In the late 1970s and early 1980s, the world's computer technology innovations advanced rapidly, and the computer science education changed with each passing day. The American Association for Computing Machinery first introduced the "ACM78 Teaching Plan", which put forward specific requirements for the purpose of teaching computer science, the curriculum, teaching materials, practice links and teachers. A few years later, the Institute of Electrical and Electronics Engineers also introduced the "IEEE/CS83 recommended teaching plan".

The computer teaching team of the NUDT actively followed up the international trend, learned from world experience and combined the teaching and research practices of the university's tasks. The teachers formed a curriculum and teaching system with its own characteristics and international standards and compiled a series of teaching materials with national influence. For example, Chen Huowang, Sun Yongqiang and Qian Jiahua jointly wrote a textbook on the principles of computer compiling that was used by nearly 200 universities. The database textbook written by Zheng Ruozhong was the first book that introduced databases in China; the numerical analysis textbook written by Qi Zhichang bundled numerical algorithms, programming and applications, which can be considered as a model of the integration of theory with practice. The data structure textbook written by Wang Guangfang has become one of the three most influential data structure textbooks in China; the text Discrete mathematics, written by Wang Bingshan and others, was concise and rigorous. The depth and width of its contents had reached the level of first-rate university science teaching materials. Guo Haozhi and others compiled program design materials and related them closely to the compiler principles, and carefully organized and published computer software practice tutorials, which had been widely circulated in domestic universities. They also used the summer school to invite teachers of computer science from other colleges and institutes and conducted in-depth discussions on the principles of compiler theory, operating systems, data structures, discrete mathematics and program design. These measures had improved the teaching levels, promoted the construction of computer courses, expanded the impact of computer education in colleges and created conditions for the construction of postgraduate courses and graduate student training.

In 1981, when the state implemented the regulations of academic degrees, computer architecture became the first batch of doctoral programmes and it was one of the five key disciplines in the country. Soon, computer software also became a doctoral station, and the computer department became one of the first three postdoctoral research stations in the country. In 1988, the computing and application major ranked the second in the national assessment of teaching.

During the period of continuous innovation in computer education, the third-generation research elites represented by Yang Xuejun, Liao Xiangke, Song Junqiang, Wang Zhiying, Zhang Minxuan, Liu Guangming and Xu Weiji stood out from the "Yinhe project". These researchers laid a solid foundation and accumulated power for sprinting.

In 1989, China launched the first outstanding teaching achievement awards. The computer teaching team of the NUDT had summed up in-depth the traditions and advantages of the discipline construction and formed a teaching philosophy based on the principle of "persisting teaching, scientific research and production, strengthening the construction of computer disciplines and actively adapting to the needs of army building". Chen Huowang, Li Yong and Peng Xinyu were recommended on behalf of the entire department and won the highest award of the nation's first outstanding teaching achievement—the grand prize.

The recognition of the country and the people's rewards gave them the power to move forward. The teaching team took the trophy and immediately began to increase the pace of innovation.

In the undergraduate professional catalogue released by the Ministry of Education in the 1980s, the computer science discipline has two majors, computer applications and computer software. They believed that the scope of computer applications was very wide and should include computer software. From the perspective of talent cultivation, the two should be combined. In fact, students who had learned both hardware and software showed weaknesses on the other one after working. On the other hand, international computer teaching programmes such as "ACM78" or "IEEE/CS83" were not divided into "hard" and "soft" professional divisions, but were divided according to science and engineering. "ACM78" is a computer science recommendation programme, and "IEEE/CS83" is a computer engineering recommendation programme. Therefore, they were thinking of combining two majors into one, which was computer science and technology, and implemented wide scope cultivation.

The reform idea received full support from the university. Therefore, they began to merge two majors into one starting in

1989, and they focused on the knowledge and ability structure that the computer professionals should master. They designed six series of courses including circuits, computer mathematics, programming languages, computer hardware, computer software and computer applications. They wanted to cultivate students' knowledge and skills in six areas by optimizing the series of courses and deepening the reforms of the teaching content, strengthening the construction of the teaching team and the construction of textbooks and laboratory conditions.

The NUDT was the first university in the world to merge the computer hardware and software into one major. In 1991, the Association for Computing Machinery (ACM) and the Institute of Electrical and Electronics Engineers Computer Society (IEEE CS) jointly launched the "CC1991 recommended teaching plan" integrating "hard" and "soft" together. In 1992, the Ministry of Education also merged computers and applications with computer software, and enrolled and cultivated according to computer science and technology.

In 1993, the country organized the second session of the Outstanding Teaching Achievement Award. The "Strengthening the Construction of the Computer Discipline and Actively Adapting to the Needs of Army Construction" plan, which reflected the content of teaching reform of the computer teaching team in the NUDT in the late 1980s and early 1990s, again won the first prize of the National Outstanding Teaching Achievements Award.

The achievements of the major teaching reform directly cultivated the growth of the fourth generation of Yinhe, including Lu Kai, Yang Canqun, Zhu Yutong, Xiao Liqun and Zhu Xiaoqian. They became the backbone of the Yinhe projects, necessary to make Yinhe products brilliant.

The computer teaching team of the NUDT continued the improvement project. In the late 20th century and the beginning of the 21st century, on the basis of the previous round of reforms to realize the "combination of left and right" disciplines, they followed the international teaching reference plans newly introduced by the ACM and IEEE and analysed in-depth the professional curriculum systems of first-rate universities and famous military academies at home and abroad. They continued to increase the interlinkages and connections between all levels of personnel cultivation to form a complete curriculum system, which is a curriculum design covering the bachelor's, master's, and doctorate degrees and dividing courses into the categories of the professional core curriculum, specialty-oriented courses and professional development courses. They also built 10 series of subject courses, including computer hardware basics, computer system architecture, computer hardware applications, programming, computer theory, software engineering, system software, network and communications, artificial intelligence and graph and image processing. They strove to achieve four characteristics, including the development of outstanding ability, highlighting military characteristics, attaching importance to theoretical foundations and striving for overall optimization.

"The Innovation and Construction of the Computer Science and Technology Curriculum System", which included the guiding principles of this education reform and its practical achievements, won the first prize in the National Outstanding Teaching Achievement Awards in 2005 out of more than 1,800 national participating projects. The award indicated that the computer teaching level of the NUDT had achieved a new leap and also a new breakthrough in the research and practice of general disciplines concerning how to both follow the trends of scientific development and how to meet the special needs of cultivating military talents.

With the ongoing development of supercomputer technology

and the continuous advancement of computer teaching reforms, the Yinhe projects continued to be successful. The research team was growing and maturing, and it formed a new scientific research camp that described the third generation of Yinhe researchers as "leading the team", the fourth generation as "commanding in the frontier" and the fifth generation as "charging forward".

In this camp, there were six scholars from the Chinese Academy of Sciences and Chinese Academy of Engineering, two members from the Academic Degrees Committee of the State Council, two national outstanding teachers, one outstanding backbone teacher of the national colleges and universities, five outstanding teachers from the army, five gold award winners of the Yucai Award, one outstanding national scientific and technical worker, 10 young and middle-aged experts with prominent contributions for the country, seven winners of the Fok Ying Tung Education Fund Award, one winner of the "Ho Leung Ho Lee Foundation" of the Science and Technology Achievement Awards, three winners of the "Ho Leung Ho Lee Foundation" of the Science and Technology Progress Awards, five first-class winners of the Guanghua Science and Technology Fund, five winners of the National Science Fund for Distinguished Young Scholars, two winners of the "Qiusi" Practical

Engineering Award for Distinguished Young Scholars, four winners of the National Young Scientist Award, three winners of Professional Technical Contribution Awards of the army, four first-class merit gainers and more than 60 second-class merit awardees. Five people were selected into the first and second level of the national "Talents Project" and six were selected for the "New Century (Trans-Century) Talent Support Program". The group also included three specially appointed professors as "Chang Jiang Scholars", 64 individuals who enjoyed government allowances and 15 professors who were above technical level three. Over 70% of the teachers had doctorates. The high-performance computing innovation team won a prize in the First Scientific and Technological Innovation Group Awards of the army, and was selected together with the high-performance microprocessor technology innovation team into the programme "Changjiang Scholars and Innovative Research Team in the University" led by the Ministry of Education.

Through the rapid advancement of scientific research, the development and expansion of the talented team and by continuously expanding the discipline space and raising the level of disciplines, the team created a high-level interdisciplinary system that supported the leap-forward development of high-performance computers. It covered five disciplines, including engineering, science, military science, network security and cryptography. It had six first-level disciplines, including computer science and technology, electronic science and technology, biomedical engineering, system science, mathematics and science of the armed forces, and 10 second-level disciplines, including computer system architecture, computer software and theory, computer application, information security, bioinformatics, microelectronics and solid electronics, system analysis and integration, computational mathematics, applicable mathematics and cryptography principles and practice. It opened up more than 20 avenues for research, covering high-performance computer architecture, high-performance microprocessor architecture, high-performance interconnected communications, military computer architecture, computer science theory, software engineering, system software, distributed computing software, large-scale parallel application software, massive data processing and artificial intelligence, virtual reality and human-computer interaction technology, network-based computer application technology, network security supervision and security operating system, biological information processing and application, military CPU technology, military DSP and

SOC, VLSI design theory and technology, reliable technology of microelectronic devices and circuits, biological information processing, biomedical imaging and image processing, complex system modelling, simulation and evaluation and system security. In 2007, computer architecture, computer software and theory and computer application technology were rated as key national disciplines. Microelectronics and solid electronics were rated as key national disciplines for cultivation, and cryptology was rated as a key discipline in Hunan Province. The first-level discipline of computer science and technology was always among the best ones in the list of comprehensive evaluations of national first-level disciplines. The interdisciplinary system covered all technical demands for developing supercomputers.

In addition, in the early 20th century, under the support of "211 project" and "985 project", the university built a modern high-performance computing innovation platform, a high-performance microprocessor technology innovation platform, a high-performance computing application research centre and a high-performance computing network support centre. The Yinhe team had already bid farewell to the difficult days of trying to create miracles out of limited materials.

What was particularly valuable was that the NUDT's College of Computer Science had a unique spiritual core. It was the "Yinhe spirit", which was formed by the wisdom, sweat, painstaking effort and even lives of the first and second generation of Yinhe researchers who had "passion for the country, solidarity and cooperation, aiming for the peak and fighting bravely". With Zhang Jinghua, Yu Tongxing, Liu Shien, Liu Qiaoyi, Zhou Jianshe and Liu Xueming as the secretaries of the Party Committee and political commissars of the College, each generation of political workers all regarded the "Yinhe spirit" as a family heirloom, to be cherished and inherited, and which should be carried forward as each new era contributed to it. It persisted in using the "Yinhe spirit" to motivate the team, to pool strength and guide innovation, which created a scientific research team that is under the Party's command, can fight a hard battle, has a good style and a strong will and is invincible.

By obtaining such a large number of high-level scientific research achievements and possessing so many technical talents, creating such a wide-fronted, deep-depth, high-level system of disciplines, building such advanced environmental conditions for innovation and acquiring a precious spiritual advantage, the overall level of supercomputer research team reached such a high level that it is not only unique in China, but also outstanding in the world.

China's supercomputing technology has already acquired the favourable climatic, geographic and human conditions necessary for reaching the summit.

Chapter VIII
Standing on the Top

In the face of a decisive opportunity and with great powers of supercomputing, the supercomputer technology innovation team from the NUDT exclaimed with pride and confidence, "For the important task of charging the peak of world supercomputer technology, we are uniquely capable!"

After having successfully developed the first phase of the "Tianhe-1" system, which was China's first supercomputer with a speed capable of petaflops and which was ranked the fifth in the world, they bravely chose the road not yet taken and successfully developed the second phase of the "Tianhe-1" system in one stroke, by continuous and unstinting effort after the first victory, winning recognition as the number one supercomputer in the world with overwhelming superiority. For the first time since the Opium War, the Chinese people stood on the top of world in scientific and technological competition!

I. The Road to Surpassing the Competition

From the PLA Military Institute of Engineering to the National University of Defence Technology, the team working on computer technology innovation had always been like the combat forces which dashed towards the frontline and fought at the forefront, having created a series of victories in the field of Chinese computer technology:

In the late 1950s, the first special-purpose vacuum tube numerical computer was successfully developed in China;

In the mid-1960s, the first general-purpose transistor computer was successfully developed in China;

In the late 1960s, the first integrated circuit computer was successfully developed in China;

In the mid-1970s, the first megaflops-capable computer was successfully developed in China;

In the early 1980s, the first 100 megaflops supercomputer was successfully developed in China;

In the early 1990s, the first gigaflops supercomputer was successfully developed in China;

In the late 1990s, the first 10 gigaflops supercomputer was successfully developed in China.

…

In the 21st century, this team, which is always ready for a hard battle and capable of winning the battle, exclaimed again, saying valiantly, "In the decisive battle with other computer great powers, if we cannot do it, who can?"

With the rapid development of national comprehensive strength, the demand for high-performance computing had become increasingly urgent in the modernization of the national economy, defence and army. The construction of a series of supercomputing centres were being planned at the national level, in provinces (cities), academies (institutes) and at famous universities. China's supercomputing cause had entered an unparalleled period of flourishment. The computer technology innovation team of the NUDT, with Yang Xuejun as its head, believed that, confronted with the rare opportunity of charging the world technology Qomolangma and the extremely urgent national demand for supercomputing technology, the Yinhe researchers should establish themselves through decisive victories and strive to render meritorious service to the country and make great contributions!

In 2006, China launched the construction of a high-performance computer and network service environment, and formulated a two-step strategy involving the development of two supercomputers of 100 teraflops for the first step, and one supercomputer of petaflops for the second step.

With the famous computer manufacturers of Sugon and Lenovo starting to develop the 100 teraflops supercomputers of "Sugon 500A" and "Shenteng 7000", respectively, the research team from the NUDT aimed to develop a supercomputer of petaflops and launched a fierce sprint towards their goal, instead of waiting around aimlessly or relying passively on missions assigned from their superiors.

The 17th CPC National Congress, which was held in 2007, pointed out that China must improve its capability of independent innovation, build an innovation-oriented country, speed up the establishment of national innovation systems, and master core technologies in the information industry as the key point of the development of Chinese scientific and technological strategies. The congress also formulated the strategic decision of the establishment and improvement of the scientific research and manufacturing system for weapons with military-civilian integration and incorporation of the military into the civilian and army talent cultivation system and army supporting system, exploring a military-civilian integration road with Chinese characteristics.

At the end of that year, then General Secretary of CPC Hu Jintao proposed during his visit in Tianjin that the Binhai New District should strive to take the lead in the whole country in regard to not only the application and implementation of the scientific outlook on development, and the improvement of a sound and rapid development in society, but that Binhai New District should also take the lead to ensure the improvement of people's livelihood and to facilitate social harmony, becoming a pacesetter in the thorough application and implementation of the scientific outlook on development.

In order to implement the call of the Party Central Committee and General Secretary of CPC Hu Jintao, the Tianjin Municipal Party Committee and the municipality proposed that Binhai New District would give full play to its function of motivation, demonstration and service. It would function as a gateway and would provide guidance with scientific and technological innovation, support the high-end industry, promote service ability and guarantee environmental development, taking the initiative of exploring a development-oriented scientific and harmonious road.

At the same time, the Party Committee of the NUDT realized during an in-depth study of the spirit of the 17th CPC National Congress that the school ought to make greater contributions in the construction of an innovative country, take the lead in military-civilian integration, and provide more technical and talent support in the promotion of local economic development, because of its strong ability to run schools, its experiences in a large number of scientific research fields for national defence with quite a lot scientific achievements and strong advantages in science, technology and talents.

With the same goal and common aspirations, the NUDT and Binhai New District in Tianjin were closely linked together. In February 2007, the two parties signed a comprehensive technological cooperation agreement through amicable negotiation.

Focusing on the national major strategic demands, by means of full development of the policy and resource advantages from Binhai New District as a national comprehensive reform pilot zone and the technology and talent advantages from the NUDT, the two parties aimed to expand mutual cooperation, accelerate the construction of bases for scientific and technological innovation and achievement transformation, actively strive for significant science and technology projects and international frontier projects, greatly enhance the ability of independent innovation, speed up the transformation of scientific and technological achievements and achieve mutual benefits and win-win results, realizing a common development in the joint promotion of an innovation-oriented country and an informationized army.

The two parties thought that the supercomputing technology was a high-tech and cutting-edge technology related to national security and development, and that it was also an important symbol of a country's strength in terms of its economy, defence capabilities and technology. They agreed to concentrate the advantageous resources of two units to make a contribution in seizing the commanding height of strategic technology with respect to supercomputers.

Therefore, the NUDT and the Tianjin municipal government jointly launched the construction project of the "National Supercomputing Center" in Binhai. This centre would be built to be a research and development centre for national high-performance computing applications, a large-scale integrated circuit centre and a basic software engineering centre. It would achieve this by jointly undertaking the development of the petaflops high-performance computing system, finally realizing a "three-in-one" information industry cluster which incorporates supercomputing services, technological research and development and personnel cultivation.

Soon after, the innovative cooperation between the NUDT and Binhai New District in Tianjin received great support from the country. The national "863 Program" included the petaflops high-performance computing system as one of its major special projects. In the meantime, the high-performance general-purpose microprocessor and high-end server operating system were also listed as major special projects in the national "Core Electronic Devices, High-end Generic Chips and Basic Software Project".

After the attack target was finally specified, the attack route was the next key point to be determined.

Two years prior, when Yang Xuejun had been in a discussion with his team over "Imagine", a 64-bit stream processor, another kind of electronic chip with the subtlety of the same function had been also dwelling in his mind, i.e., the scientific computing of

Graphics Processing Unit (GPU). That is to say, in the development of a petaflops supercomputer, the established heterogeneous parallel technology route for the CPU+64-bit stream processor could proceed, and another heterogeneous technology route for a CPU (general-purpose microprocessor) + GPU (special-purpose microprocessor) could also be tried.

For the former, the team had abundant technical accumulations with many years of special research, which accordingly had a higher probability of success. For the latter route, the highest computational efficiency of the GPU was generally considered to be only 20%, which was in no way useful in supercomputer development.

However, the advantage of building a supercomputer with a GPU is obvious: with its high computing speed, six times higher than that of the CPU, the machine space could be reduced effectively; its low energy consumption, which is only one fifth of a CPU, could efficiently make up for the weakness of high energy consumption in the supercomputer; its many varieties were circulating on the market, providing more choices for the mature technology; and its low price could significantly improve the cost performance of machines and render them affordable to customers.

The disadvantages of developing a supercomputer with GPU

technology could not be ignored either. GPU applicability in high-performance computing was still unknown and the road of exploration was full of hardships and risks.

Yang Xuejun, as the chief designer of the research team, decided to make a bold trial of CPU+GPU heterogeneous parallel architecture after in-depth technical investigation and research, together with repetitive weighing of the advantages and disadvantages.

Abruptly leaping to one petaflops from 10 teraflops, the decision to take the CPU+GPU technological route, which had never been taken by others, triggered an uproar in the industry.

"The development of a supercomputer normally proceeds at the increasing pace of 10 times, which has been an international practice. Isn't it too rash a step while directly leaping over the 10 teraflops to the one petaflops? Can this be leaped over?"

"Even if the machine can be manufactured, can its application level be enough to fulfil realistic demands?"

"As the applicability of GPUs in high-performance computing is still a mystery, isn't it too risky to develop a supercomputer with it?"

...

In face of tumultuous doubts, the research group led by Yang Xuejun thought that even though the opportunity to simplify two steps into one step seldom arose in the world, it was not the first time for Yinhe researchers. In those years when Ci Yungui led everyone to develop the central computer for the "Yuanwang 1" tracking ship, was it not also a direct leap from 10 kiloflops to one megaflops? In today's world, the supercomputer is improved in performance by 1,000 times every 10 years. In such a circumstance, we will always be behind if we still follow the so-called practice of step-by-step development, aping others at every step. We can only fight our way out and take the lead by way of finding a new path. As for the application possibility of image-processing GPUs in scientific computing, Yang Xuejun confirmedly indicated that the team could branch out by analogy and conquer the difficulty of GPU computing performance with the help of decades of knowledge accumulated in supercomputer teaching and scientific research, together with innovative achievements attained in the research of the 64-bit stream processor.

The Party Committee of the NUDT actively supported their heroic undertaking and issued a mobilization order, noting that "the size of an aspiration that is cherished in our mind determines how splendid our cause will be", encouraging everyone to establish their determination for a decisive engagement and to strengthen their confidence of victory.

The leading group of the college, with Zou Peng as the dean and Zhou Jian as the political commissar, in consideration of the demands of task, compiled an “aircraft carrier fleet” of supercomputer innovation through an organic organization of innovative teams from the college in the fields of high-performance computing, high-performance microprocessing, basic software, network technology and application technology.

With the great trust of the country and national expectations on the shoulders, the “aircraft carrier fleet” of supercomputer innovation from the NUDT was advancing in a formidable array toward the new technological destination!

II. The Breakthroughs

The heterogeneous parallel system of CPU+GPU, figuratively speaking, is a "strap-on rocket" that links organically various CPUs and GPUs (a CPU is the equivalent to the core and the GPU functions as the boosters).

According to this principle, the research team ingeniously divided the supercomputer system into computer arrays, acceleration arrays, and service arrays, to improve the computational efficiency to the utmost extent, to reduce energy consumption and costs, and to accelerate the speed through the heterogeneous cooperation of CPUs and GPUs.

The most innovative point of this technological route lay in the application of the image-processing GPU in high-performance computing, the biggest challenge of which was to ensure the high-efficiency computing of GPUs. This problem had been the first barrier to hinder the advancement of petaflops supercomputers.

At the end of 2008, the research team, with Yang Xuejun as the chief designer, assigned this important problem to a small contingent of researchers, led by Yang Canqun.

After more than 10 years of practical experience in scientific research, Yang Canqun described the work with a very ingenious metaphor, "Engineering and technological activities are just like playing a guessing game. When the riddle is dissolved, everyone will see the light, saying, 'It has not turned out to be that recondite'. But before that, amidst thick clouds and fogs, you do not know where to head and do not have any clue of which direction to take, aimlessly idling around."

Scientific calculation using GPUs was just such a riddle.

At the time, it was claimed that there were two kinds of GPUs with general-purpose computing capabilities, which were manufactured by NVIDIA and AMD with multiple models each. A separate GPU was just a chip that was applicable only in the form of a graphics card, which is composed of matched memory and peripheral circuits. There were several graphics card manufacturers and nearly 20 types of graphics card available on the market. Among these numerous graphics cards, which one could meet the scientific computing requirements? Yang Canqun and his team members felt themselves lost in the darkness.

Since the beginning of 2009, the group led by Yang Canqun started the screening job day and night with the purpose of finding a graphics card which had high computing performance in terms of double-precision floating-point arithmetic, nice system compatibility and stable operation.

One week before the Spring Festival, they installed a type of graphics card on a mainboard for test, but the system failed to start after the installation of the software system. They thought there was a hardware problem at first, but the hardware technicians insisted that the mainboard was of high quality. They searched for causes in the software without any solution to the problem, even though they had tried different versions of operating systems and graphics drivers. They worked extra shifts during the Spring Festival in order to find the underlying causes of problem. On the fourth day of the lunar new year, they found an indistinct marking on the mainboard, saying that there had been a malfunction over an abnormal start-up, and it was not confirmed if the malfunction had been thoroughly solved after reparation, which amused and saddened everyone.

There was also a graphics card with two GPU chips, the drive program of which required two monitors to allow the two GPUs to work simultaneously. This obviously could not meet the requirements for scientific computing, because it was impossible to install a large number of monitors in a computer system. They found by consulting data that it was possible to simulate a monitor by connecting a resistor to the output of the graphics card. They searched out several plug-in resistors that could meet the requirements from a components cabinet that was covered with the dust of more than 10 years, and finally solved the testing problem.

...

During the course of two months, after going through innumerable twists and turns, they completed the testing and installation of nearly 20 types of GPUs and finally found a GPU that met the computing requirements.

What would be the compatibility effect of a CPU+GPU heterogeneous parallel system, which combined thousands of CPUs and thousands of GPUs together?

In March 2009, they combined a CPU and a GPU and conducted

the evaluation by using the GPU acceleration program, and found that the overall performance was less than 60 gigaflops. Even a single CPU could achieve 50 gigaflops. In other words, the computing efficiency of the GPU was still only about 20% when it was combined with CPU for scientific computing, even though the GPU had remarkable speed in image processing.

Everyone's heart was chilled with disappointment after this test result. Then one question had to be asked: Was the idea that a supercomputer of petaflops could be manufactured with GPUs of such a working efficiency just a fairy tale? Was it as foreign experts had asserted, that GPUs could not be used for scientific computing?

The chief designer Yang Xuejun rushed to the laboratory the first time after he had received the report, and firmly said, "The road which others did not dare to take is not surely to be a dead end. It can be concluded from the analysis of technical principles that the computing performance of GPU can be substantially enhanced through software optimization."

Zhou Jianshe also came to laboratory to encouraged everyone, saying, "We should foster the 'Yinhe spirit' of courageous action and tenacious striving, break through all difficulties and obstacles to create new glory for the 'Yinhe' projects!"

The university then held a project mobilization meeting for tackling the petaflops supercomputer. The university and college Party Committee made a call to everyone, saying, "We should shoulder the important historic mission of charging against the world technological summit of a supercomputer for the country and nation, and resolutely complete the research task of a petaflops supercomputer at the end of this year. Let the song of triumph play around the whole nation and the splendour of Yinhe radiate again all over the world!"

At this critical moment, then CPC Secretary General Hu Jintao nominated Liao Xiangke as the Dean of the College of Computer Science upon recommendation of the university Party Committee. Liao was also appointed as the chief commander and standing deputy chief designer of the project at the same time.

Liao was a brilliant leader who had matured during the development of high-performance supercomputers. He had successively participated in the development of the five generations of the Yinhe high-performance computers, the "Yinhe Kylin" operating system, the security operation system of an information processing platform and a confidential server and movable parallel computer. He served as the supervising designer, director designer, deputy chief designer and standing deputy chief designer. When he was responsible for the research of parallel processing software of the operating system in the development of the "Yinhe-2", he designed the two-level scheduling algorithm of the "task-logical CPU-physical CPU" by applying the concept of the "logical CPU", which had facilitated the parallel efficiency of the "Yinhe-2" to reach the leading level in the world. While he was the director designer of the I/O operating system in the development of the "Yinhe-3", he emphatically solved the technical difficulty of equalization between I/O capability and computing capability of the massive parallel processing supercomputer. Meanwhile, he was also responsible for the research of the "Yinhe-3" distributed main memory parallel mechanism, and proposed a user-level communication optimization protocol which significantly improved the practical communication performance of the massive parallel processing system. The domestic "Yinhe Kylin" operating system, which had been developed under his leadership, was the server operating system with the highest security level in China at that time. Characterized by high performance, strong security, nice availability, efficient real-time functioning and expandability, this system finally passed the acceptance inspection in December 2006, and passed the structural protection level assessment by the Ministry of Public Security and the military safety certification organization. This was the first general-purpose software which had passed the equipment design finalizing examination. It then was listed among the 10 major scientific and technological breakthroughs of China's colleges and universities and received high praise from central leaders and experts in the national 10th Five-Year Plan major scientific and technological achievements exhibition.

In the face of setbacks, the general designer Yang Xuejun and chief commander Liao Xiangke unexpectedly made an abnormal decision. They decided to shift the timeline to an earlier release date and bring out China's first supercomputer of petaflops before the end of 2009, one year ahead of the formerly scheduled time of 2010.

When the decision was announced, some people were taken aback, asking with wide eyes, "Is it feasible to complete the task a year in advance when the critical techniques have not been tackled yet?"

Nevertheless, the new generation of Yinhe researchers were confident in themselves, saying "Wasn't there a huge difficulty when we developed the "Yinhe-1" in those years? But the seniors indomitably advanced and completed the task a year ahead of time, beyond everyone's expectation. The "Yinhe-3", which had also been planned to be completed within five years, was actually realized within only three years with a huge stride from one gigaflops to 10 gigaflops by feat of concerted efforts and race against the time. What the seniors have achieved will be repeated in our hands!"

Under the leadership of Yang Xuejun and Liao Xiangke, the research team from the NUDT kicked off the general offensive against the supercomputer of petaflops.

On the bank of the Xiang River in the northern suburbs of Changsha, a field was embraced by mountains. Up on the mountain side there were grass and trees, lush and green, and down the mountains there was a three-story small building. This was the location of the Changsha flood control headquarters. As it was not yet the flood season, this quiet and secluded place was sparsely populated, with only birds chirping here and there.

Yang Canqun and his team regarded this place as a battlefield. Every day, they isolated themselves in the building from the outside world with only one concern in the mind, which was how to give play to the role of GPUs. The only place their eyes were staring was the display screen, searching for any fleeting inspiration and capturing, one after another, opportunities to optimize the computing efficiency of GPUs from tremendous amounts of continuously rolling data, and subsequently revising computing programs again and again.

During that whole week, Yang Canqun and his partners were staring as usual the screen from 7 a.m. to midnight and from Monday to Friday without any chance for success or any achievement.

After several days of engagement in the fierce battle, utterly worn out for quite a few days, Yang Canqun lay in bed, tossing about all night without getting a wink of sleep. He did not reconcile himself to the realistic situation. Normally, some breakthroughs for performance optimization could always be found from Monday to Friday, and then they could take the weekend to study optimization methods. The data were scuttling to and from in front of his eyes like a swarm of bees, and rolling in the mind like billowing waves when they closed their eyes.

Suddenly, he faintly felt that some of the data scrolling in front of the eyes were lower than the design objective. He rolled out of the bed, rushed to the office from home, opened the laptop which was connected with the server, entered the test database, and indeed found that some of the GPU computing resources had not been used yet. Finding it hard to suppress the excitement, Yang Canqun immediately set about the program optimization and succeeded in improving the computing performance of GPUs once again. When he rose to his feet and opened the door after completing the modification of the program, he saw that the sun had already climbed to the top of mountains, smiling brilliantly, and birds were dancing jubilantly in the woods, warbling melodiously.

They carried out similar optimizations and improvements like this more than 10,000 times, and finally promoted the GPU computing efficiency to 58%.

This fully testified that the CPU+GPU heterogeneous parallel system was scientific and feasible!

Yang Canqun and his team continued their triumphant advance through more victories, repeatedly conducting tests and discussing and striving for improvement. Although each improvement was as tiny as a drop of water, the accumulation of them could create scientific miracles. After four months of consecutive hard work and more than 80,000 instances of improvement and optimization, the computing efficiency of the GPU finally jumped to the highest level of 70%!

III. Charging in All Directions

The American computer genius Seymour Cray said, “It’s easy to manufacture a high-speed CPU, but difficult a high-speed system.”

The quote from “the father of supercomputing” was fulfilled again in the development of China’s first supercomputer capable of petaflops.

In July 2009, they built a system of several cabinets in light of the CPU+GPU architecture, only to find that the stable running time of the system could hardly exceed half an hour during trial runs. How could this be?

Everyone thought that the problem lied in the GPU, after preliminary analysis. Besides the computational efficiency of

GPUs in scientific computing, there was another relevant and very important factor which was the stability of the GPU. The GPU was designed for graphics processing, and thus its computing load was quite different from general-purpose computing. Especially after the overall performance of GPUs had been brought into play, all of the components were under a heavy load status with increasing power consumption, stringent heat dissipation requirements and degraded stability of all devices. When more GPUs were installed in the system, the mean free error time of system would be accordingly decreased.

Without a solution to this problem, the road of CPU+GPU architecture would remain a dead end.

The issue of how to improve the working stability of the GPU became once again a new riddle to be urgently resolved by Yang Canqun and his team.

First of all, they conducted a pressure test of all the GPUs, one by one, through the screening method in order to find those GPUs which ran smoothly and stably. However, the result was not satisfactory. The system stability had been improved, to an extent, but it was still far away from the design requirements.

Surrounded by vast and dense fogs, they seemed to be trapped within a bewildering maze, ignorant of the direction ahead and the way out. They always believed that the day would finally dawn, regardless of the long and endless night.

Being still absorbed in the arduous exploration, the unit had a dinner party to celebrate Army Day on 1 August. Yang Canqun said to his comrades, "Come on! Let's go get a drink. Break time." When they were at the table, no one really remembered what they had drunk or eaten. What occupied their mind was still the working stability of the GPU, even when they were having drinks and food. After putting down their bowls and chopsticks, they returned to the computer room and buried themselves among the vast GPU technological data, racking their brains for that glimmering light of hope that would break through the dense mist, eating and sleeping continuously in the computer room for several days.

On the morning of 4 August, a post about how to improve the performance of GPUs by overclocking, which he had browsed on the Internet, suddenly came to the mind of Yang Canqun. According to the post, GPU performance could be improved by means of overclocking, but it would also cause the unstable running of the GPU and might even case a system blackout.

Yang Canqun suddenly had the strange inspiration that if the selected GPU possessed the function of frequency modulation, wasn't it possible to improve its stability by underclocking the GPU?

As luck would have it, the GPU which was picked out and installed possessed exactly the function of frequency modulation. Everyone immediately carried out the underclocking test. As a result, the stability problem of GPU was readily solved.

The exploration of GPU computing efficiency and its critical technology of stability was full of difficulties and reverses, and there were also ups and downs and heart-gripping risks with each step in tackling other key technologies.

To reach the speed of one petaflops required not only fast-computing CPUs and GPUs, but also a fast and smooth network system which enabled all kinds of information to be fast-transmitting.

In October 2008, Su Jinshu led a team to undertake a development project of a new type of switch, an "overpass" for the communication network of the petaflops supercomputer, which directly determined the data transmission speed. Upon in-depth investigation and research, and strict argumentation, they proposed the orthogonal interconnection scheme, which could simplify the system structure and reduce the design difficulty, the requirements for the manufacturing process and costs for development and production.

However, after an exchange with an American technical engineer from a chip manufacturer about the scheme of orthogonal interconnection via the Internet, he strongly opposed this scheme and emphasized repeatedly the following viewpoints in three emails and four telephone meetings.

Firstly, they had also studied the orthogonal interconnection and conducted simulations, experiments, and tests under related conditions with the conclusion that it could not meet the signal transmission requirements of this type of switch due to large signal transmission loss and discontinuity of impedance.

Secondly, no one could be successful without their technical support, because the development of this type of switch took them more than two years due to its complication in design.

Thirdly, they would not provide the technical support if the Chinese team insisted on the scheme of orthogonal interconnection.

Fourthly, the insistence on the scheme of orthogonal interconnection would be destined to be a failure.

The research and development work was caught in a dilemma from the very beginning. Once they insisted on their own design route, everything needed to be explored all over again with great difficulty in design and high risks. If they changed the scheme and completely proceeded according to the scheme provided by the Americans with some certainty of success, they would not have their own features, neither in innovation nor advantages.

The Yinhe researchers, who had never followed and worshiped others blindly and slavishly, unhesitatingly made the former choice. Through over two months of simulations and experiments performed day and night, they really found that the test scheme and design specifications of the Americans could not fulfil the signal transmission requirements of the new switch under the condition of orthogonal interconnection. Whereas, after in-depth study of the relevant theory and techniques with respect to signal integrity and the design specifications of the United States, they finally found that the rectangular antipads that the Americans used were the main cause of large signal transmission loss and discontinuity of transmission impedance. In view of this weak link, they invented the racetrack-shaped and dumbbell-shaped antipads and obtained the overall design specifications through repeated iterative simulation over three months. The technical parameters of the critical eye diagram had reached 60ps, which was far greater than the 35ps of the Americans.

They had completed the development of the new type of switch within only 10 months, the actually measured technical indicators of which greatly surpassed equivalent systems with only 80% of the costs of equivalent and similar-size products.

Quite a number of setbacks were also experienced during the tests of the interconnection chips and interface chips of the high-speed network.

When the development team had completed the logic design and hardware and software simulation through hard work for more than half a year, there were only three days left before the deadline for tape-out.

All the final versions of the logic designs were integrated in the FPGA test software for the last full test. This was also one of the key points for development work. If it was passed, everything would advance smoothly; if not, everything would be wasted for nothing.

When the program was being started up, everyone stared at the screen with wide eyes. All of a sudden, all the data, which had been merrily rolling on the screen, were motionless as if they had hit a wall.

Everyone's heart sank and they asked each other, bewildered, what the problem could be.

They quickly inspected thc periphery only to find that the fibre was intact; they checked the server and it was also active. They examined the switch and it was also charged; they went over the

Ethernet and it was also smoothly running.

In the end, there was nothing to try but a reboot. The data began rolling for several minutes and then reverted to their original state, lying there motionlessly.

Deadlock! Everyone was in a cold sweat from anxiety. The source of the malfunction must be found out if they wanted to get past this problem.

The connection point between the test questions and test mode was initially selected as the potential problem in checking the causes. Liu Lu, who was responsible for test operation, and Xie Min, in charge of the design test questions, were high-minded with youth and impatient from anxiety, and quarrelled immediately when they met each other.

Busy with testing another driver, Xie Min questioned Liu Lu face to face, asking “Why are you asking for me when it is clear that I’m busy?”

Liu Lu retorted, “When we integrated all the test questions into the system, it ran for a while and stopped with deadlock. Isn’t it possible that the test questions you have programmed are not compatible?”

Xie Min continued, “Impossible! There will be no impact from mixed questions if a single question runs smoothly."

Liu Lu added, "Are you sure? Running with a single question and running with mixed questions—can that be the same?"

Xie Min continued, "You can rest assured that I have conducted the flow control internally for all questions. There is no problem with them!"

Liu Lu inquired, "Now that it runs smoothly with other questions independently, it indicates that there is no problem with the hardware. If it's not because of your questions, why does the deadlock happen only when your questions are running?"

Xie Min retorted, "I suspect that there is problem in your test mode!"

Silent for a while, they waved at each other almost simultaneously, "Let's stop wasting time in quarrelling." They agreed to check their own causes to justify themselves and discuss together to find the solution, according to the established practice.

All the personnel led by the director of the research laboratory had been busy with checking for the cause of the fault all day long, only to find that there was no problem in either test questions or the test mode.

Could it be a problem with the interface chips? After checking the values on the statistical counters of all the interface chips, it was confirmed that the quantity of data packages flowing out of four interface chips was exactly the same with that flowing into the four interface chips, which signified a normal result.

They had no other choice but to transfer their suspicion to the interconnection chips. The consequence would be unbearable to consider if the interconnection chips were found to be the cause at the last moment. When everyone was busy with reading codes, in great suspense and agitation, some doubtful points were found accidentally in the testing FPGA (Field-Programmable Gate Array) version.

After updating and rerunning the FPGA version, the data were running again merrily. Finally, a successful tape-out of both of the chips had been conducted.

A series of critical technologies were tackled one by one through arduous and persistent efforts.

One was the "Yinhe Kylin" operating system. In consideration of the realistic demand for a petaflops supercomputer, the research team of the basic software developed a new version of the "Yinhe Kylin" operating system on the basis of a transformation and update of the original version. The new "Kylin" operating supported the heterogeneous parallel system, 64-bit multi-core and multithreading microprocessor and SoC architecture. It also supported high-speed interconnection communication based on advanced routes and supplied multi-level parallel compilation optimization support with the capability of high-performance virtual computing domain management and a realized integrated energy consumption management framework. It was an authentic product "made in China", an operating system with the highest security level of that time.

Another project was the high-speed interconnection communication technology based on advanced routes. It broke through the technology barrier of on-chip advanced network system architecture, and independently designed high-efficiency communication protocols and an advanced tile-type on-chip interconnection network and high-density inter-chip Internet, which enabled the bandwidth of link bidirectional communication to reach 160 Gbps and single backboard switching density to reach 61.44 Tbps, two times and 2.37 times, respectively, those of the international mainstream commercial interconnection IB QDR.

Another feat was the development of multi-level parallel compilation optimization. An easy-to-use and efficient application programming and operation could be achieved by the conduction of a multi-core and multithreading scheduling mechanism, a multi-level parallel dynamic load balancing algorithm, interprocedural analysis of the total system and other compiling algorithms, which efficiently supported the JASMIN programming framework.

The team also made great strides in the high-performance virtual computing domain. It broke through the technologies of high-efficiency user containers, load balancing and virtualized network terminals, and innovatively realized the functions of security isolation and customizable user environments in high-performance computer systems and efficiently enhanced the security and usability.

Finally, the team managed to develop integrated low consumed power control based on software and hardware technology. An integrated energy consumption management framework was designed to effectively reduce the power consumption in a system running through self-feedback cooling regulation of the monitoring system, processor frequency and voltage regulation, self-adaption node energy consumption status transitions and other methods.

…

On the eve of National Day in 2009, the installation of the first phase of the petaflops supercomputer was completed. At the same time, another piece of good news was coming from the chip market that a new GPU with higher performance had entered the market.

Everyone was overjoyed by this news, as the Yinhe researchers were still competing fiercely with world great powers. It felt like an open net appearing in front of the forward in an international football match.

However, taking the shot was not that easy. The first problem was that only one month was left before the scheduled task node deadline. Second, there were a total of 2,560 nodes in the system. It would cost half a month for the team to disassemble and reinstall all the GPUs. The original software optimization must also be improved after the replacement with new GPUs. Could this task be completed on time?

The final shock was approaching, along with the National Day. Testing, screening, disassembling and installing GPUs constitutes hard manual labour. All of the members of the team, regardless of age and sex, finally completed the replacement of thousands of GPUs after three days and nights without getting a wink of sleep.

With these tasks having been completed, Yang Xuejun once again called everyone together. His eyes were moist when he caught sight of everyone's hands, which were covered in Band-Aids, and their red eyes.

IV. Go Out of Asia

The sudden advent of China's first petaflops supercomputer made China the first country in the world to master the CUP+GPU heterogeneous parallel architecture and the second country worldwide to successfully develop a supercomputer with a computing speed of petaflops.

On hearing the news, then General Secretary Hu Jintao inscribed in person the words "Tianhe" as the name of the computer.

In the late autumn of 2009, Changsha, the provincial capital of Hunan Province, was enjoying fine weather. The sky was blue, the water was crystal clear, the mountain was shrouded by a canopy of red leaves, and golden fruits were hanging heavily on the branches.

In the harvest season, Hans Meuer, the founder of the international TOP500, came to Changsha with his testing crew and carried out an actual measurement for the performance of the "Tianhe-1" supercomputer in the NUDT.

As the founder of the international TOP500 list for supercomputers, Hans Meuer has a mind of wide scope. At the moment he set his foot on the floor of the facility room of the "Tianhe-1", a pair of careful eyes could still notice a change in the expression on his face, as indicated by his raised thick eyebrows.

The "Tianhe-1" supercomputer system displayed in front of Hans Meuer was really breath-taking. In the facility room of nearly 1,000 square meters, rows of exquisite cabinets were erected like proud squadrons in a military parade to be reviewed and inspected. Hundreds of thousands of indicator lights flashed here and there as if in a green Milky way on the Earth.

But it was the "Tianhe-1"'s unique technologies and outstanding performance that Hans Meuer felt more stunned by.

The "Tianhe-1" worked with a peak performance of 1.206 petaflops, although the actual measurement of performance by LINPACK was 563.1 teraflops. This meant that the computing capacity of the "Tianhe-1" in one day is tantamount to that of a microcomputer equipped with an Intel dual-core CPU working at a main frequency of 2.5 GHz for 160 years!

The shared memory capacity of the "Tianhe-1" amounted to 1PB in total. Calculated according to the PDG format of the domestic digital book application software, if every book was 10MB on average, the memory capacity of the "Tianhe-1" was equivalent to four national libraries with a collection of 27 million books each, and was capable of storing a photo of everyone in the country of approximately 1MB.

The "Tianhe-1" had a transmission rate of 10 Gbps for a single network cable, which was the fastest worldwide at that time and was as good as running in an information highway built inside the "Tianhe-1" computer.

The "Tianhe-1" consumed 1,280 kW of electricity per hour in operation, which meant a computing performance of 430 megaflops per watt, an advanced level worldwide.

...

At that time, authoritative experts who were participating in the academic annual meeting of the High-Performance Computer China in Changsha said after obtaining actual measurement data of the "Tianhe-1" from experts of the international TOP500, "The invention of the 'Tianhe-1' marks a leap forward in China's research and development capabilities in supercomputing from the level of 100 teraflops to that of petaflops, which is of important strategic significance to solve challenging difficulties in China's economy, science, military and other sectors and is also important for the drive of building China into an innovative country and the promotion of its overall strengths."

On 18 November, the International Supercomputing Conference was held in Portland, a western city of the United States. At the moment when the international TOP500 released the 34th TOP500 List, exclamations of surprise echoed at the conference.

With the list came two pieces of unexpected news. One was that Cray, from America, the "godfather" in the field of supercomputing, had finally taken the place of IBM, a champion on the list for many years in a row, in the top spot, for its "Jaguar" which operated at a peak speed of 2.331 petaflops and with an actual measurement performance of 1.759 petaflops. Another was that the "Tianhe-1" had been listed in the top five worldwide, the best place that Chinese computers had as yet achieved on the TOP500 List.

It was praiseworthy and very unusual for the "Tianhe-1" to achieve this ranking among the top five. Among the top 10 computers, nine were developed by America and only the "Tianhe-1" was a product from China. This was the first time that the Chinese computers were crowned as the best in Asia, becoming the "King of Asia" in the field of supercomputing. Computers on the list came from all five continents. The top spots in the other four continents were taken by machines unsurprisingly made in America, but the "Number One in Asia" was the only exception, on the cup for which the words "Made in China" were distinctly carved.

Hans Meuer, the founder of the international TOP500, said at the conference that it was amazing and stunning that scientists from the NUDT in China had successfully developed the "Tianhe-1" through the heterogeneous parallel technology of CPU+GUP!

After the ceremony, Hans Meuer said once again to the Chinese representatives at this conference that it was by no means a simple task for a Chinese computer to be listed among the top of a list which is usually monopolized by American computers.

Professor Wang Baosheng from the NUDT walked onto the podium with delight and, while receiving the cups for the "Global Top Five" and the "Top 1 in Asia", was so excited that he almost forgot his award acceptance speech.

Off the podium, Wang Baosheng, still lost in excitement, added:

"It felt like winning a medal in the Olympic Games. It would be perfect if the national anthem of the People's Republic of China was performed just like in the Olympic Games."

V. In Pursuit of a Greater Victory

Receiving the call of congratulation across the Pacific Ocean, Yang Xuejun, the chief designer of the "Tianhe-1", murmured an "oh" with a smile and then put down the mobile phone. Since his joining the Yinhe crew, particularly since the moment when he had assumed responsibility as the chief designer for the research and development of Yinhe, he had led the crew to crack difficulties again and again in the forefront of the supercomputing sector and had reaped various satisfactory achievements which had been awarded the grand prize and the first prize of the National Defense Science and Technology Progress Awards, the first prize of the National Teaching Achievements Awards, the second prize of the State Technological

Invention Awards, the Professional Technical Contribution Awards of the Army, the award of the National Science Fund for Distinguished Young Scholars, the Foundation of Innovative Research Groups and the a first-class merit. He just smiled at the news of congratulation or the certificate of an award or medal and let it go every time that he won an award. As far as he was concerned, awards are tantamount to "just having finished a thing again", as his mother says if he has completed a project, overcome a difficulty and reaped an achievement, no matter how influential.

As a veteran fighter on the forefront of the field of supercomputing for many years, Yang Xuejun was quite well aware of the fact that the solid support of scientific technologies was in urgent need for China's rapid development, on the one hand, while, on the other hand, China was by no means comparable with developed countries in terms of research and development capabilities of supercomputing, not to mention its application abilities, which lag far behind. In the fierce arena of competition of high-performance computers, a slight relaxation would lead to an inferior situation or even remove China from the competition entirely. Thus, the successful invention of the "Tianhe-1", like the scientific and technological barriers he had each time broken through, was simply another milestone. There were many more problems to solve.

One evening, Yang Xuejun, the chief designer of the "Tianhe-1" project, Liao Xiangke, the chief commander of the "Tianhe-1" project, and Zhou Jianshe, the Political Commissar of the College of Computer Science, took a walk in the square in front of the office building on thc campus.

Yang Xuejun said, "The Central Committee of the CPC has proposed a grand target to build China into an innovative country and to realize the informatization of the PLA. Playing a vanguard role in the course of rejuvenating the country through strengthening the army, universities will assume long-term heavy missions."

Liao Xiangke added, "According to the comrades attending the award ceremony of the international TOP500, although we are fast advancing in the ranking, there is still a huge gap between China and the developed countries. Among the entire TOP500, 277 systems come from America, while only 21 are made in China. China has made an overall plan to overtake the international advanced technologies of supercomputing. As the national team of innovative technologies in computer science, we must play our role as much as possible."

Yang Xuejun replied, saying, "Although we are honoured with

the top spot in Asia, we must not put our focus only on Asia, but should go global."

Zhou Jianshe said, "To strive for the top and surpass the rest of the world is the dream the Yinhe team has been pursuing generation after generation for several decades. We were all extremely excited the moment we heard the news that our computer had soared into the top five. With passion in our heart to drive us forward, all of us are eager to achieve a greater victory in the second phase of the 'Tianhe-1' project and climb to the top of the Mount Qomolangma of the field of supercomputing."

Yang Xuejun said, "In the second phase of the 'Tianhe-1' project, besides achieving an overall promotion in every aspect of the performance of our computer, we must apply the CPU developed by ourselves to the computer and gradually get rid of our dependence on imported microprocessors."

The "Chinese computer with a foreign core" was an untold regret and a hidden wound in the heart of the Yinhe team.

In an effort to equip the Chinese computers with a Chinese core, the microprocessor technical innovation group of the NUDT started to design and develop the "Feiteng-1000" (FT-1000) chip when the "Tianhe-1" project was initiated in 2008.

In order to make the FT-1000 chip achieve an international advanced level and to facilitate its application in a wider scope as well as a sustained development, the research team, in line with the tendency in the development of CPUs, chose the SPARC instruction system, which is excellent in comparability, and adopted the multi-core and multi-thread SoC architecture. There are eight processing cores, each of which supports eight threads, in the chip, because of which the FT-1000 has become the processor with the most threads in a single chip in China. In addition, in order to meet the demands in the research and development of supercomputers, a three-path direct chip interface has been integrated within the FT-1000, with the help of which a multipath SMP system can be formed through a direct interconnection between two to four processing chips. The 4 MB sharing second-level cache and the four-path DDR3 memory controller (MCU) are also integrated to better match the processing of data with the memory access bandwidth and to ease the pressure of the memory wall.

Someone vividly described the research goals as "soaring up to the sky in a single bound". This phrase indicates an imposing vigour as well as harsh obstacles.

Shortly after the research and development commenced, the

testing of the DDR3 (Double-Data-Rate Three) was confronted with dual challenges. There was an inconsistency between the control chips on the DIMM (Dual-Inline-Memory-Modules) with the latest DDR3, which had led to the loss of data because the instructions could not be stored when data of various ranks were refreshed simultaneously, and, due to the large scale of the chips, it was difficult to encapsulate them well, which resulted in an undesirable duty ratio of the chips. Through arduous hard work for several nights in a row, the problem was satisfactorily solved after the best solution had been found and the optimum scheme had been decided at last.

But the manufactured sample was unqualified in performance. After tremendous effort, they noticed that there was a problem in the top-down design due to an underestimation of the difficulties in the top-down system by their cooperating companies. There was no way out but to come up with a new design from scratch. A new physical design must be determined to increase the performance of the product to a large extent.

In October, due to the massive design workload, the synopsis ICC tool couldn't function and it was impossible for the Cadence Encounter compiler to achieve a basic routing. It was only two months before the tape-out deadline.

Everybody was quiet, aware that the more urgent a situation is, the more they must stay calm. After a careful analysis of the data for design and a sort-out of the data flow direction, a new top-down design was proposed. Though this new proposal involved a rework of the top-down design as well as a series of designs associated with the power consumption and the encapsulation and was heavily-tasked, it was scientifically feasible. Thus, both the research team and the cooperating companies were willing to offer close collaboration. Through more than 20 days of tremendous effort to finish this urgent task, the path timing violations had been finally fixed.

The time flew by and it was again deep winter. This was the time for the last inspection of the chip design—the inspection of timing sequence. An unexpected problem arose just when everybody thought they could go home to have a sweet sleep after they had experienced countless sleepless nights to find measures to overcome various difficulties—there were major hidden problems in the design procedure regarding the layered delay computation and the integrity of the signal. The design of the whole system of the CPU would be a sheer failure if such issues couldn't be eliminated.

Therefore, all the crew members restarted with high morale and carried out intensive inspections to find solutions. In the end, the causes were successfully wiped out and normal data were obtained again.

The design of the general-purpose CPU for the FT-1000 was completed on time, with successful tape-out in one attempt.

On the next day, the technical innovation team of the College of Computer Science of the NUDT held a meeting to encourage and inspire the team staff working on the second phase of the "Tianhe-1" project. Raising their right hands high up, all the members yelled out a loud, resounding slogan like their predecessors of Yinhe projects:

In a timeline of no more than a year!

No less than 4.7 petaflops!

The self-developed Feiteng CPU must be applied to the computer!

VI. The Peak

After hearing about their determination, many experts expressed their deep admiration. However, they were also concerned, saying, "Unless there will be a miracle, to improve the performance of the machine by nearly three times within a year is impossible."

It was not just a simple change in numbers from the 1,206 teraflops of the first phase's system to the 4,700 teraflops of the second phase's system. The peak value of computation would be increased by nearly three times, but the number of cabinets could be only increased by about one fourth, which meant that, with a same-sized cabinet, the second phase's system would be two times better in performance than that of the first phase. This had brought development work with a great many challenges. There were lots of technical problems, including design techniques for multicore/multithreading architecture and on-chip parallel technology. Meanwhile, compiler optimization, an autonomous and highly efficient communication protocol, advanced router architecture, and the design of an ultra-large-scale IC and high-speed/high-density exchanger were also problems that needed to be solved. None of these technical problems were minor. Instead, every one of them was a dreaded obstacle that only tremendous effort could help overcome.

The installation of fibre cables was the primary project, since the second phase's system of the "Tianhe-1" had moved to the National Supercomputer Center in Tianjin (NSCC-TJ). Time was pressing, and the task was arduous. In order to guarantee the completion of this project before the deadline, the director had a daily schedule prepared, specifying tasks for each day. He asked everyone to not eat or sleep until they had finished their daily task.

Yang Canqun led the research team to start the development work at NSCC-TJ. Their first task was to ensure that all the components in the system would run continuously and steadily for over four hours. However, problems arose the moment they turned on the computer.

Before Yang and his team went to Tianjin, they had done system verifications for four cabinets and conducted stability tests without discovering any problems. The components that NSCC-TJ used were exactly the same as what the research centre in Changsha used. Why was there a problem?

Yang, looking up at the Tianhe computer room, felt that they had a long way to go. There were 140 cabinets standing in parallel, including tens of thousands of components. If a single component or a single system went wrong, the stability of the whole system would be affected. Which component or system created the problem? There was only darkness ahead, as if Yang and his team had entered a deep hole.

Exploring in the darkness for a few days, they finally discovered that the water cooling system was the problem: inadequate water volume and decreased diffusing capacity had caused the overheating of the supercomputer's system.

As system debugging was in full swing, they noticed that there were some unexpected undulating phenomena in the GPUs. Everyone had done a great many sampling analyses, including the analysis of the GPU itself, the power supply module, the GPU and the mainframe's communication interface adapters, the GPU's heat dissipation system, and so on. However, they could not find anything that might explain the strange undulating phenomena. Again, they monitored the GPU's temperature in the operating state. With a great amount of data analysis, they found that there was an obvious temperature difference between two GPUs which were on the same position in the operating state. Through the ventilation technique, the problem of heat dissipation was finally solved, which maintained the GPU's stability.

The second phase's system of the "Tianhe-1" used a self-developed interconnection network system, which was also a key element that would affect the system's stability in the operating state. As the machine itself was enormous and its structure was quite complex, it was fairly difficult to test it. When something came up, searching for the cause and performing later maintenance were both challenging. By cooperating with scientific researchers of the interconnection network system, Yang and his team wrote a variety of test codes for grouping and paralleling, which efficiently brought about the full coverage of the network interface and the network path. In this way, fast fault locating and troubleshooting were achieved.

The second phase's system of the "Tianhe-1" was on track for the current release schedule before the coming National Day.

Yang and his colleagues, who had already worked on this project for two months without rest, had no time to sit down and have a cup of tea. They immediately optimized the system's performance one last time. They tested every single computational node in the system one by one. In this way, they troubleshot problems that would affect the performance, such as memory failure and GPU faults, which enhanced the performance to 1,890 teraflops.

When this task was finished, they took advantage of this favourable situation and again optimized the application software, improving the system performance to 2,339 teraflops.

This was already a miracle. The American "Jaguar" supercomputer, which topped the list of the world's supercomputers at the time, could only run at 1,767 teraflops. Thus, the performance of the "Jaguar" was far behind that of the second phase's system of the "Tianhe-1" based on the evaluation of the system's performance on the TOP500 list.

But Yang and his colleagues were not satisfied. They believed that the "Tianhe-1" still had some hidden potential that they could explore. The better the "Tianhe-1" could be in performance compared to the "Jaguar" supercomputer, the greater the impact the "Tianhe-1" would exert on the world.

Yang and his colleagues stayed in the computer room, making

the final dash to achieve their goal.

On 19 October, Yang went to the Beijing office in the afternoon. The car galloped along the Jingjin Expressway. When it went over an overpass, he saw that cars coming from all sides gathered here and drove away in all directions. An inspiration came to his mind: if the interconnection network was an urban traffic hub, the network paths would have been all the urban streets. The intersections of these streets were often the places where traffic jams took place. Only when these cars were allowed to pass in a proper way would the problem of traffic congestion be solved, and, in this way, smooth traffic could be ensured.

Yang called his colleagues immediately. He asked them to pay attention to the network path, make some changes in the parameters, and optimize the performance of the supercomputer one more time.

That night, the performance of the "Tianhe-1" went up to 2,490 teraflops.

The next day, a miracle happened once again—2,507 teraflops!

On 30 October, the eve of submitting the test result to the TOP500, they still kept optimizing the system and successfully improved the performance to 2,566 teraflops and increased the calculating efficiency to 54.6%, which was the world's most advanced level.

The scientists who took part in this project kept working selflessly, overcoming scientific challenges one after another. They came up with a solution to a difficult problem worldwide, which was the high-speed and high-efficiency interconnection between supercomputers' CPUs. Meanwhile, an advanced interconnection chip and high-performance interface chip had been successfully developed. The researchers' achievements also included four types of node machine, two kinds of networks and 15 printed circuit boards. They had developed a specific operating system, compiler, parallel programming environment and visualization system of scientific computation. Among them, the heterogeneous parallel architecture system and high-speed network communication technology, which was based on energy-efficiency routing, had met the international leading level.

They also installed 2,048 general-purposed "Feiteng1000" chips on the second phase's system of the "Tianhe" and achieved a breakthrough in terms of the development of the "Chinese core". A complete domestic production could be achieved if users asked for it. Through high-efficiency interconnection and communication, a self-developed high-performance supercomputer could be produced.

One of the main authors of the TOP500 list was Jack Dongarra, a computer science professor at the University of Tennessee. After studying the second phase's system of the "Tianhe-1," he commented, "Although the processors of the 'Tianhe-1' were still American products, its interconnection chips were completely made by China. Also, China has had its own CPUs. The interconnection chips were mainly involved with information communications between processors, which had a key impact on the supercomputer's performance. The production of such chips had reached the world's most advanced level."

Professor Dongarra was a well-known expert in the field of international high-performance computers, so his comment was rather objective. The advanced chips in the routers and the high-speed network chips developed by the NUDT were two times better in performance than the imported commercial chips. The successful use of the "Feiteng1000" chips on the "Tianhe-1" represented the end of the period without domestic chips in the Chinese information industry.

Compared with the first phase's system of the "Tianhe-1", the performance of the second phase's system was significantly improved once again. The peak value of its speed achieved 4,700 teraflops and the sustained speed reached 2,566 teraflops, an increase of 2.89 times and 3.55 times, respectively. The calculating efficiency was improved by nearly 10%.

In November 2010, with its peak value of calculating over two times higher than that of the "Jaguar" supercomputer, the second phase's system of the "Tianhe-1" was ranked the first among the world's TOP500 supercomputers at the Supercomputing Conference. At the same time, the "Sugon Nebulae" supercomputer, which was deployed at the National Supercomputer Center in Shenzhen (NSCC-SZ), won third place. In addition, 39 supercomputing systems produced domestically in China were in the TOP500 list as well. The Chinese supercomputers accounted for 8.2% of the world's TOP500, two times higher than a year earlier.

The fact that the "Tianhe-1" had topped the TOP500 list and that the global occupancy rate of the Chinese supercomputing systems rose at a fast pace ended America's long-standing dominance of supercomputing technologies. This victory not only demonstrated that China's self-developed supercomputing technologies had reached the world's leading level, but also demonstrated a rapid enhancement in the Chinese information industry, scientific and technological innovation ability and overall national power.

Liu Guangming, professor at the NUDT and the Director of NSCC-TJ, received the award on behalf of the research team of the "Tianhe-1". It was a golden medal with the inscription "Made in China".

This "scientific Olympics" did not have the glorious stadium of a real Olympics. There were no national flags or national anthems. However, this award did not only represent a single person, a few people, or a dozen people's wisdom and techniques. It symbolized a nation's comprehensive strength. Compared with the Olympic gold medal, this award was far more significant and would have a far-reaching impact on the world.

Chapter IX Championship and Dominant Strength

The world's supercomputer powers were not happy about losing their dominant status in regard to supercomputing technology. They launched a series of counterattacks and regained their dominant status, initiating a "tug-of-war" of supercomputers in the world.

The supercomputer research team of the NUDT successfully launched the "Tianhe-2" and again ranked the first place among the international TOP500, retaking the crown and repeating the feat for five consecutive championships.

China supercomputing had become the superhero to revitalize the Chinese nation.

I. Tug-of-War for the Championship

On 31 October 2010, the UN Secretary General Ban Ki-moon visited Nanjing University. In his speeches to the teachers and students, he said, "Every time I come to China I marvel at its dynamism and the breath-taking speed at which it is changing. I saw this today on the new Huning High Speed Railway from Shanghai. Three hundred kilometres in just over an hour. This is not even your fastest train. Last week, China introduced the world's fastest scheduled service between Shanghai and Hangzhou. And I read in the news that China is a front-runner to build a super-fast computer. China is, indeed, a country on the move."

The honest and friendly scientists in the international community were all happy for the success of the "Tianhe-1". Thomas Sterling, a professor of computer science at Louisiana State University, said: "There is no doubt that today the whole world is paying attention to China and its technological development due to the emergence of the 'Tianhe-1'. It is predicted that the supercomputer field will become an area in which China will develop rapidly in the future. We look forward to more exciting achievements."

Jack Dongarra, one of the international TOP500 initiators, was very optimistic about the future development of China's supercomputers, saying "The 'Tianhe-1' is 40% faster than the supercomputer developed by the United States' Oak Ridge National Laboratory, which is in fact a great improvement of computing speed. Although there is still a gap between China and the United States in the aspect of processor technology, China is committed to the research and development of this area at this moment. Therefore, I will not be surprised if China achieves the same advanced level of processors as the United States in one or two years."

A Canadian online commenter said, "If the 'Tianhe-1' represents a technological leap for China, then we are likely to see more leaps in its technology exports. When we have children, we must let our children learn Chinese in order to be prepared when Chinese becomes the latest trendy language."

Of course, the United States had the strongest reaction to the leap in development of China's supercomputer technology.

When a senior computer expert from the Virginia Polytechnic Institute in the United States heard the news of the "Tianhe-1" winning the championship, he said to reporters, "The emergence of the second-phase system of the 'Tianhe-1' in China was unexpected and we were caught off guard. The United States has not yet been prepared psychologically. We can predict that there will be many unexpected surprises happening in the future of China. The United States must be ideologically prepared as early as possible."

On 3 October 2010, President Barack Obama bluntly stated at the first press conference after the midterm election of the House of Representatives, "Nowadays we should have the consensus that theoretically China should not have a railway system more advanced than ours and Singapore should not have airport better than ours, but I just heard that China now has the fastest supercomputer in the world!"

After analysis and comparison, one Hong Kong media outlet reported, "In the past 100 years, very few countries have shocked the United States as the 'Tianhe-1' did with any technological invention."

Why would the Americans be so shocked? Obama himself

explained the reason, saying "We always rank the first in this field." Now, China had taken this first place, indicating that China was investing in their infrastructure and expecting long-term returns from these investments. Based on such logical reasoning, it is easy to see that China will create "challenges and threats to the future of the United States."

Although this logic has the suspicion of "subjectivism", it is not unreasonable.

Supercomputer technology, as one of the three pillars and powerful engines supporting and leading the development of current science and technology, is also the strategic advantage by which the United States maintains its technological leadership, especially in the aspects of advanced weaponry and equipment. It is also an outpost for controlling the world and dominating international competition. Therefore, the United States had always guarded this area carefully and never conceded.

Let us first look at all the number one supercomputers since the establishment of the International Top 500 from 1993:

June 1993–November 1993: the "CM-5", manufactured by an American company;

November 1993–June 1994: the "Numerical Wind Tunnel",

manufactured by a Japanese company;

June 1994–November 1994: the "Paragon XP/S140", manufactured by an American company;

November 1994–June 1996: the "Numerical Wind Tunnel", manufactured by a Japanese company;

June 1996–June 1997: the "SR2201" and the "CP-PACS", manufactured by a Japanese company;

June 1997–November 2000: the "ASCI Red", manufactured by an American company;

November 2000–June 2002: the "ASCI White", manufactured by an American company;

June 2002–November 2004: the "Earth Simulator", manufactured by a Japanese company;

November 2004–June 2008: the "Blue Gene/L", developed by an American company;

June 2008–November 2009: the "Roadrunner", made by an American company;

November 2009–November 2010: the "Jaguar", manufactured by an American company;

November 2010–June 2011: the "Tianhe 1", developed by the National University of Defense Technology of China.

…

From the above rankings, it can be seen that the American-made machines had won the most champions and had occupied top position for the longest time. At the same time, the vast majority of the top 10 machines were made in the United States and they hold a 50% share of the TOP500, which was indeed half of the world. In the field of supercomputers, other than the United States, only Japan was sometimes able to show its competitiveness. However, the United States seemed reluctant to accept this.

Take the "Earth Simulator" as an example. In June 2002, Japan launched the "Earth Simulator" for the exploration of Earth physics. It replaced the American "ASCI White", winning the top ranking. This supercomputer was more than five times faster than the "ASCI White," and its computing power was even stronger than the sum of the top 20 machines in the United States.

"The reaction of the United States to the emergence of the 'Earth Simulator' was just like its reaction to the Soviet Union when the Soviet Union successfully launched a man-made Earth satellite in 1957", as the international TOP500 founder Jack Tongala described the shock that the "Earth Simulator" brought to the United States.

The United States acted quickly. On one hand, it invested

heavily to implement a supercomputer strategy to surpass other countries. On the other hand, the United States. changed its focus of supercomputer applications from the development of weaponry and equipment, which had been implemented since Cold War period, to all sectors of society. Thus, the United States initiated the "general war" of supercomputers.

In order to persuade the United States President to approve this plan and secure sufficient funds, the National Science Advisory Committee listed various challenging issues from nine areas in the report, such as climate prediction, transportation, bioinformatics and computational biology, social health and safety, earthquake predictions, geophysical exploration and Earth science, astrophysics, materials science and computational nanotechnology and human organizational system research.

The American media called this plan the "Supercomputer Crusades".

Two and a half years later, the "Blue Gene/L" developed by the United States ended the myth of the "Earth Simulator" and the researchers proudly declared, "Japan, as a 'dark horse' in the supercomputer technology field, has promoted the United States to improve the level of its supercomputers."

The United States unable to accept Japan ranking the first in the supercomputer technology field, and it considered Japan to be its follower, ally and pawn. How would the United States see the potential threat of China's supercomputer technology?

Steven Chu, the United States Secretary of Energy, delivered a speech at the United States National Press Club, saying "Only last month, China's NUDT developed the fastest computer in the world, which presents a challenge to us. The United States should take action to promote innovation, on which we always do the best!"

The Wall Street Journal published an article regarding this, saying that while China is achieving the installation of the fastest computer in the world, the government should take action to restore its leadership in this field and accelerate the current most powerful computer by 1,000 times its computing speed in order to surpass Chinese engineers and exceed Chinese computers. China plans to develop a fully autonomous and innovative microprocessor and apply it as the core engine for computers in the future. If that day comes, China will have less and less reliance on American companies in regard to microprocessors, thus strengthening its resistance to exported products from the United States.

Yomiuri Shimbun quoted the words from Professor Matsuoka

of the Tokyo Institute of Technology, saying, "Compared to the rapid decline of Japan's advantage in the supercomputer field, China has only just started to take off. It is only a matter of time before Japan's supercomputer technology is overtaken by China."

Chosun Ilbo (Korea Daily) from South Korea also expressed worries in their reports, "The supercomputer research and development in South Korea had been interrupted and we must re-promote this work. Supercomputers are strategic products with a view to the future in 20 years and are closely related to national dignity. This is the similar to the launching of satellites and space aircraft. The United States and China have already erupted into a computer war which is no longer an ordinary war. Therefore, we have to think about the development direction of South Korea."

The United States launched a transcendental plan to fight back, Japan began to implement its strategy of revitalization and Europe did not want to lag behind and so followed this trend as well.

The new round of "tug-of-war" of supercomputers began in this context.

II. Setting Out Again Quietly, Based on Past Peak Achievement

Just six months later, when the international TOP500 released a new ranking list in June 2011, the "K Computer", installed in Rikagaku Kenkyusho (Institute of Physical and Chemical Research) and developed by a Japanese company took the initiative and replaced the "Tianhe-1", ranking first in the TOP500. In June and November of 2012, the American supercomputers the "Sequoia" and "Titan" once again achieved the first place of the international TOP500, successively. The "Tianhe-1" then fell to number eight.

This ranking fall of the "Tianhe-1" made the fans of domestic supercomputers so distressed, heartbroken, and disappointed.

“What happened to the ‘Tianhe-1’?”, they asked, “How come it was surpassed so quickly by the development tide of supercomputers from other countries?”

Meanwhile, some people with dubious motives began to clamour again, saying “Domestic machines are like this; they are just a political specimen.”

At that time, Tianhe research personnel were surprisingly calm, not panicking, not explaining, not refuting, and even not frustrated.

Tianhe research staff had early predicted that Japan and the United States would overtake China again. Supercomputer technology was always their area of advantage and strategic field, where the United States and Japan had been showing their strength and standing above others. How could they allow a “dark horse” to continually exceed them? Moreover, surpassing and being surpassed, as well as being looked up to and being looked down upon are both normal states of the development of science and technology and the driving force of technological advancement. There is no need to worry about it or panic about it. Instead, silence best reflects self-confidence and strength.

More importantly, although the “Tianhe-1” was successfully crowned in the number one spot of the TOP500 with applause and flowers, making Chinese people proud and gratified, Tianhe research personnel never became complacent or impatient. They were fully aware that the rules of the game in the world's supercomputing field would never change because of the appearance of the "Tianhe-1."

Let us have a look at what Tianhe staff said to media reporters.

"As far as overall strength is concerned, the United States is still standing in the first echelon. The temporary victory of the 'Tianhe-1' can only indicate that we are ranking top in the second echelon."

"In the latest TOP500 rankings, the United States has more than 230 computers on the list, and all of them are developed by American companies. In all, 409 computers out of the top 500 are developed and manufactured by HP, IBM, and Cray. IBM employees are circulating the following joke internally: 'In the field of supercomputers, 97% of the market share comes from IBM, and the remaining 3% comes from the second-hand machines of IBM'. Japan has 30 computers on that ranking list, among which only 37% were made by Japanese companies and the rest were made by American companies. China has 76 computers ranked in that list, but only 13% of these 76 computers were made by China. Most users in the telecommunications and the Internet use HP and IBM systems. Compared with the United States, the overall level of China's supercomputers does not just lag behind a little but a lot."

"The development of the overall system of computers in China has already taken its place in the front ranks of the world, but there is still a long way for China to go in terms of a complete industrial chain of high-performance computers."

"Architecture, interconnecting technology, operating systems, microprocessors and application software are the five indispensable core requirements for supercomputers. China has handled the first three very well, but the latter two are still the weak points of Chinese supercomputers."

"In regard to core components and original technology, there is still a big gap between the level of China and the advanced level of foreign countries. For example, the physical design of CPUs has a gap of at least one generation compared with those of the United States, while the manufacturing technique has at least two generations of difference compared with that of the United States."

"In regard to the application of computer technology, it is the same situation as above. In the countries with an advanced level of supercomputing technology such as the United States, Japan and some other countries, supercomputing has achieved deep integration with social production and development, greatly promoting the rapid development of a large number of other industries such as automobiles, aircraft, aerospace and movies. However, the supercomputers of China are only successfully applied in some of these technical industries, resulting in the application bottleneck which has not yet been completely broken through, not only affecting social progress but also delaying the development of supercomputers."

"In regard to computer talents, China is even less competitive. The United States has more than 10,000 senior supercomputer professionals, while China is unable to find enough talents even with high salaries. NSCC-SZ even found it difficult to recruit high-qualified talents with an annual salary of RMB 1 million."

"Although the 'Tianhe-1' won the champion among the international TOP500, the dominant position of the Western countries in the field of information technology is not changed; the dominant position of the United States in the development and application of supercomputers is not changed, and the situation that the world powers compete for the dominant position of supercomputers is also not changed."

The above three statements, that things were "not changed", were an accurate summary of the strengths of various countries in the supercomputing field, vividly reflecting the calmness and peaceful mind of Tianhe research people after they ranked the first.

Gaps accumulate energy, and distance stimulates power. Yang Xuejun, an academic of the Chinese Academy of Sciences and also the chief designer of the "Tianhe-1", said, "From the first day of the 'Tianhe-1''s emergence, we already started our research on the 'Tianhe-2'. After analysing the developing trend of international high-performance computers, we aim to develop supercomputers that can achieve 10 petaflops. We are very determined to make some new contributions to lead the development of supercomputers in the world."

"Eating one dish in the bowl while watching the next dish in the pot and thinking about other dishes that haven't been prepared" is the traditional thinking mode of Yinhe and Tianhe research staff.

"To take action is always better than to tell others" is always the behavioural style that Yinhe and Tianhe research people adhere to.

They had just occupied the top rank, and they were already working towards the new challenge based on this peak achievement they had made.

In January 2011, the "Tianhe Project Leading Group Meeting" was held in the NUDT, officially launching the certification and development of the “Tianhe-2”, which aimed to achieve 10 petaflops. Liao Xiangke, the Dean of the College of Computer Science, the general commander and deputy chief designer of the “Tianhe-1” project, was appointed as general commander and chief designer of the “Tianhe-2”.

In March, the NUDT and the Guangzhou municipal government began negotiations regarding cooperation and joint building of the National Supercomputer Center in Guangzhou (NSCC-GZ).

In November, the “Research and Development Project of New Generation of Tianhe Supercomputer” launched by the NUDT was reviewed and approved by experts organized by the Ministry of Science and Technology. At the same time, the NUDT signed the “Agreement to Build NSCC-GZ with Efforts from Provincial Government, Municipal Government and University” with the Guangdong provincial government, Guangzhou municipal government and Sun Yat-sen University. Thus, the research and development of the “Tianhe-2” started in an all-round way. Afterwards, the research team signed the “‘Tianhe-2’ Research and Development Contract in NSCC-GZ” with the Guangzhou municipal government. The campus of Sun Yat-sen University was selected as the location for NSCC-GZ.

In May 2012, the NUDT provided a supporting computer to NSCC-GZ in order to guide and assist the research and development work at the early stage.

…

After two and a half years of silence, the “Tianhe” supercomputer showed its power again. In June 2013, the “Tianhe” supercomputer ranked first among the international TOP500 again, winning another championship of supercomputing in the world!

The peak speed of the “Tianhe-2” was up to 54.9 petaflops, and its continuous calculation speed had reached up to 33.86 petaflops, representing a leading international level of its comprehensive technology.

Compared with the American “Titan”, the champion winner before the “Tianhe-2”, the “Tianhe-2” could calculate two times faster with 2.5 times higher computing density.

Compared with the “Tianhe-1,” the “Tianhe-2” improved its computing performance and computing density by over 10 times, and its energy efficiency ratio was also doubled, but the power consumption was only one-third of the “Tianhe-1”.

The “Tianhe-1” was able to simulate the climate change of 2,000 years ago in exploring the laws of climate change, but the “Tianhe-2” could trace back 5,000 years.

Using the current computer system that BGI Genomics has in order to conduct a genome-wide association analysis for 500 people would take about one year. However, it would only take three hours by applying the “Yinhe-2”.

It took more than one year to complete the animation rendering for the movie Avatar. If the “Tianhe-2” had been used, it could have been completed in one hour.

Moreover, researching and developing a new type of car with the traditional methods usually requires more than 100 crash tests and over two years of experiments. With the “Tianhe-2”, it only takes two months and requires only three to five crash tests.

The computing power of the “Tianhe-2” truly lives up to its name of “supercomputing”!

Those who made irresponsible remarks and looked for flaws of the “Tianhe” supercomputer finally shut their mouths for a moment.

Zhang Yunquan, a researcher at the Institute of Software of the Chinese Academy of Sciences, commented with pride, “On the road to computer architecture, China now is holding the hands of the world and taking the lead!”

Some foreign scientists have also made fair statements.

Rajeed Hazra, the Vice President of the Intel Corporation of the United States, said, “The progress of the ‘Tianhe-2’ will not only benefit China’s scientific community and industry, but will also promote the development of world supercomputer technology in decades. The ‘Tianhe-2’ and other supercomputers provide infrastructure for the growing global demand for big data processing.”

Horst Simon, the Deputy Director of the Lawrence Berkeley National Laboratory, commented, “If anyone thinks that the Chinese supercomputer is just a gimmick, then the ‘Tianhe-2’ proves that they are totally wrong.”

III. "Sprinting! Sprinting! Sprinting!"

How did Tianhe researchers feel when they ranked the top of the world again?

Celebrating, Yang Xuejun, the chief designer of the "Tianhe-1" and the President of the NUDT, Liao Xiangke, the Dean of the College of Computer Science and chief designer of the "Tianhe-2", and Liu Xueming, the Political Commissar of the College of Computer Science, said some representative words when they were toasting each other.

Yang Xuejun said, "After successfully developing the first dedicated numerical vacuum-tube computer in 1958 and becoming a base for computer research and talent training in China, the NUDT persisted in working on overcoming more world-leading technology difficulties, thus leading the continuous development of computer technology in China. The successful development of the 100 megaflops 'Yinhe-1' in 1983 especially enabled China to achieve the leap in development from the mainframe computer to the supercomputer. The 14-year research and development of the 'Yinhe-2' and 'Yinhe-3' from 1983 to 1997 effectively promoted the leap in development of Chinese supercomputers from 100 megaflops to one gigaflops and then to 10 gigaflops. Ten years after that, the computing speed of supercomputers developed from one teraflops to 30 teraflops and then to 100 teraflops successively. From 2007 to 2010, China firstly created the leading architectural technology in the world, assisting Chinese supercomputers to achieve a calculation speed leap from 100 teraflops to one petaflops and winning the top rank of the international TOP500. Then, Yinhe and Tianhe researchers finally realized their decades-old dreams. Now we have won the championship of the world again, further consolidating our position in the world supercomputing field. What do these leaps in development really mean? They mean that leaps in development of supercomputer technology are both our tradition and responsibility. Nowadays, with the rapid development of information technology, we must continue to challenge ourselves and surpass ourselves. Otherwise, we will be washed out by the world even with little negligence!"

Liao Xiangke stated, "Ranking first again is not a stop sign for innovation. When we were developing the 'Tianhe-1' and 'Tianhe-2', we did not use all of our potentialities, meaning that there is still great space for improvement of our technical ways and our team has great potential for innovation. We must and we are able to stand higher and go further!"

"After the 18th National Congress of the Communist Party of China, President Xi Jinping brought up the concepts of the 'Chinese Dream' and the 'Chinese Military Dream', which encouraged the majority of scientific and technological workers to be determined to contribute more wisdom and strength for the rise of China", said Liu Xueming. "Scientific research is like working on the battlefield, where offence was considered as the best defence. If we want to keep an invincible position on this battlefield, we must keep sprinting, sprinting, and sprinting!"

The supercomputer research team from the NUDT indeed have the strength to strive for greater achievements and to find greater glory.

"When I heard that the 'Tianhe-2' won another number one ranking in the international TOP500, I felt not surprised. It is a great thing! On the contrary, I would be surprised if it didn't rank No. 1", said a general who was in advanced studies in the Military High-Tech Training Institute of the NUDT. "The year when the high-tech class of the NUDT had just started the training classes, I was admitted to the regiment-level cadre training class. After that, I successively participated in division-level and corps-level high-tech cadre training classes. During the one-year study life in the NUDT, every day I saw people going into and coming out of the Yinhe Building and the Tianhe Building when I was doing running exercise in the morning. I found out later that those coming out of the building had just finished an overnight work in the research laboratory and those going into the building were just prepared to start their experiments in the laboratory in advance. At night, almost every room in both buildings were brightly illuminated. Every day was same like that, no matter in spring, summer, autumn or winter. I've never seen people working so hard like this."

The words of this general actually revealed the "soft power" of the supercomputer innovation team, which is the spirit of being courageous, enterprising and tenacious. At the same time, their "hard power" manifests in unique technological advantages.

As mentioned before, architecture, interconnect technology,

operating systems, microprocessors and application software are the five core requirements of supercomputers. The first three core requirements, in the words of Tianhe researchers, are "our unique techniques".

The CPU+GPU heterogeneously parallel architecture adopted by the "Tianhe-1" was an overall architecture technology that had a vital innovation significance to the traditional methods. It had the advantages of low energy consumption, low costs and a high degree of integration, thus quickly becoming an international mainstream. On this basis, the Tianhe team boldly innovated and designed a new heterogeneous multi-state architecture for the "Tianhe-2", greatly improving the computing speed of the system. Moreover, the system was extensively applied from scientific computing to big data processing and large-scale information services and some other fields.

With supercomputer systems becoming more and more complex and large-scale, interconnect technology is increasingly important, even as important as the CPU. The performance of the high-speed interconnect system of the "Tianhe-2" was twice that of the international commercial interconnect system at that time. It could link tens of thousands of microprocessors together to solve a computing problem, which overcame the worldwide problem in efficient interconnection regarding "lower performance with more microprocessors". They independently developed the two core components of the interconnection system: routers and network interfaces. If the system of a supercomputer is like a big city, the interconnection system is the road network of the city, the routers are the flyovers and the interface is the entrance and exit of the main roads. No matter how good the municipal facilities of road network are, the city still will suffer serious traffic jams if the entrances and exits of flyovers and main roads are not well designed. When they were designing the two chips, they applied a variety of innovative technologies, finally achieving an efficient and rapid data exchange.

As Professor Jack Dongarra responded to the reporter's question as to what made China's supercomputers so fast, he said, "China has independently developed internal interconnect technologies and those technologies cannot be bought from others. All of these are based on the chips, routers and interchangers developed independently by themselves. Their situation is similar to the case of Cray Inc. In addition to the contributions that Cray has made regarding integration and software, Cray also contributed with their internal interconnect technology. They applied InfiniBand technology in their internal interconnection, thus greatly increasing the efficiency of internal data transmission."

The operating system used by the "Tianhe" systems is also very distinctive. While most Chinese supercomputers use foreign operating systems, the "Tianhe" supercomputer applies the "Kylin" operating system which had been independently developed by China and is well known for its high security. This operating system makes every user of the "Tianhe" feel like they've rented a safe deposit box from the bank, with the keys and passwords all in their hands. In addition, the information in the box is not accessible to other users or even administrators. In a word, "It is more comfortable and safe for Chinese users to use self-developed operating systems."

Meanwhile, Chinese researchers are also working hard on the other two core requirements of the supercomputer, which are CPUs and application software, in order to catch up international advanced level.

It has been a long-cherished dream of Chinese scientists to equip Chinese supercomputers with "Chinese cores". Now that the NUDT had successfully developed the "FT-1000" CPU, successfully applied it to the "Tianhe-1" and partially replaced imported CPUs, their dream had finally come true. The domestically produced "FT-1500" CPU that was applied on the "Tianhe-2" accounted for one-eighth of all CPUs on the "Tianhe-2". If the user requires it, it is absolutely no problem to use domestic CPUs for the entire system.

The pinnacle of science has never been static but is always changing, developing and surpassing. Therefore, the peak is not the end of a scientist's pursuit, but a new starting point for the next sprint. It is always the lifestyle and life state of scientists to start again and surpass previous peak achievements.

As Tianhe researchers continue to move forward, the technology innovations of Chinese supercomputers continue to establish new world records.

In the 42nd International TOP500 rankings issued in November 2013, the "Tianhe-2" won the world championship once again.

In June 2014, the "Tianhe-2" achieved three consecutive championships in the international TOP500 rankings.

In November 2014, with a computing speed of 33.86 petaflops, the "Tianhe-2" won the championship for a fourth time as the fastest supercomputer in the world. In addition, the continuous computing speed of the "Tianhe-2" was nearly two times faster than that of the American "Titan", which ranked second. This was the fifth time that "Tianhe" supercomputers had won the championship of supercomputing in the world.

IV. A Double-Standard Championship

It is true that the development of domestic supercomputers, including the "Yinhe" and "Tianhe" series computers, were all dominated by the Chinese government. However, the research and development of the world's top supercomputers, such as the "Sequoia", the "Titan" and the "K Computer" were also directly funded by the American and Japanese governments. These supercomputers were jointly researched and developed by companies such as IBM, Cray, and Fujitsu, together with national scientific research units, which essentially similar to the research and development model in China. The research and development of supercomputers has always been serving to solve the major scientific issues involving national security and development and for enhancing the overall national strength. It is not driven purely by market behaviour and commercial interests, and this is true in any country.

The development of China's supercomputers has also been following the path of technology research and application. During the research and development of the "Tianhe-1", under the leadership of Song Junqiang, the supercomputer application innovation team from the NUDT actively penetrated the front line of users and visited key users and potential users from house to house, learning about their different application demands regarding scientific engineering calculations, big data processing, high throughput and high security information services. On such a basis, they concluded the technical requirements for the design of a petaflops-capable supercomputer, thus continuously optimizing the operating environment of supercomputers. At the same time, they promoted the characteristics of petaflops-capable supercomputers and guided users to learn and adopt new technologies, thus promoting the complementary and mutual promotion of computer machine design and application, and laying a good foundation for the new generation of Chinese supercomputers to be useful to and desired by users. After the "Tianhe-1" was put into use, five high-performance computer application platforms were established, such as oil exploration, bio-medicine, animation and film special effect rendering, high-end equipment manufacturing and geographic information, thus achieving a number of innovative achievements with an advanced international level.

After the launch of the "Tianhe-2" development and research, Song Junqiang led the team to carefully sort out and plan the scientific research directions in accordance with the multiple needs of polymorphic applications. They conducted a series of innovations in scientific engineering calculations, cloud service platforms of large-scale resources and big data processing, perfecting the "Tianhe-2" to be more useful and practical. In addition, through multi-level fault-tolerant design, researchers had achieved the intelligent management of large systems and achieved a means to monitor, detect, diagnose and isolate failures and problems automatically during operation. The time ratio of continuous system-wide stabilization was improved by 1.5 times compared with the "Tianhe-1", meaning that the reliability and availability of the "Tianhe-2" reached a new level.

From winning a single championship to winning consecutive championships, the "Tianhe" supercomputers were always recognized by the international TOP500. In addition, the ranking from this organization is not based on theory but is based on actual, measured performance by LINPACK, which has been implemented for more than 30 years and been considered the most recognized and the most authoritative system standard in the world. Since then, there have been other international rankings, such as HPCC, Graph500 and HPCG, which use different test programs to evaluate the application performance of supercomputers in certain aspects. In the Graph 500 actual measuring, because the "Tianhe-2" still had a large amount of space to be explored and improved, it had only some nodes tested, resulting in only sixth place in the ranking. The American "Sequoia" computer, which ranked third in the international TOP500, and the American "Titan", which ranked second in the international TOP500, did not even appear on the HPCG and Graph500 ranking list.

The issue of energy consumption is the biggest obstacle for supercomputers to continuously develop forward. Since the start of development and research of the "Yinhe" supercomputers, China had realized that this problem should be resolved in a timely manner. Chinese researchers had contributed a series of innovative technologies to solve this scientific problem. According to the Green500 ranking, which evaluates energy consumption, the "Tianhe-2", which adopted a new energy consumption control mechanism, was equivalent to the "Titan" and "Sequoia", which ranked second and third, respectively, in the international TOP500. In addition, the rank of the "Tianhe-2" was way ahead of the "K Computer" from Japan, which ranked fourth in the international TOP500. In conclusion, the "Tianhe-2" is of high efficiency and has good energy savings. Some "clowns" even compared the energy consumption of the "Tianhe-2" with other computers which were lower-graded by two levels, concluding that the "Tianhe-2" had high energy consumption. This is not only applying a double standard to the "Tianhe-2", but also inverting right and wrong, and making sensationalist claims!

Others said, "Quite a number of the microprocessors used in the 'Tianhe 2' are imported CPUs, which is not at all innovation."

One scientist responded very well, saying, "When you build a house and some of the bricks are not fired by yourself, then would you say that this house was not built by yourself?" In fact, with the quality of the "FT-1500", it was absolutely certain that researchers could apply only Chinese chips to the supercomputers. The reason why they just partially used chips developed in China was because most of the application software was imported and could only run on corresponding imported microprocessors. Therefore, domestic microprocessors could only be used in service arrays.

“The development of Chinese supercomputers attaches great importance to hardware but little importance to software. Some users of the ‘Tianhe-2’ need 10 years to write the necessary codes, and the users are from very few industries; thus the degree of Chinese supercomputers’ application is much lower than that of the developed countries like the United States”, judged some “famous commentators” about China’s supercomputers.

Do the users really need 10 years to write software? The Tianhe team staff pointed out sharply, “This is to confuse the longer development period in certain fields of application with the shorter ‘migration period’ which it is actually applied on the ‘Tianhe-2’.”

This squabble about the “Tianhe-2” echoed a scene that occurred at the award ceremony for the international TOP500 in November 2010.

On that day, when the copywriter of the international TOP500 announced that the “Tianhe-1” had ranked the first in the world, the ceremony venue suddenly became very noisy, with all participants making startled and shocked noises. Liu Guangming, the representative of the NUDT, went up to the stage to receive the award, but an American reporter gave him a look of anger and disdain before he left the award podium. This reporter couldn’t wait to stand up and questioned the copywriter of the international

TOP500, asking, "Do you think it is scientific to use LINPACK to measure the performance of computers and decide the ranking based on such standard?"

Since the establishment of the international TOP500, ranking standards had been running for more than 30 years, and no one ever questioned its scientific nature. Now that the "Tianhe-1" from China had won first place, some people (including some scientists) questioned whether or not it was scientific.

The international TOP500 institute adopted the suggestions of doubters and designed the Supercomputer High-Performance Conjugate Gradient (HPCG) ranking on the basis of extensive solicitation of industry insiders.

There is also a very vivid metaphor to describe the relationship between the Supercomputer High-Performance Conjugate Gradient (HPCG) ranking and the international TOP500 ranking: "They are like two sets of examination papers. The international TOP500 uses LINPACK, which is like a standard Examination Paper A in use for 30 years, while the High Performance Conjugate Gradient (HPCG) is a recently-introduced Examination Paper B. The former mainly examines operating speed, while the latter mainly measures application performance."

In November 2014, the international TOP500 organization first released the 44th list of the world's top 500 supercomputers. The "Tianhe-2" won its fourth consecutive championship. The next day, the international TOP500 organization officially released the Supercomputer High Performance Conjugate Gradient (HPCG) ranking list for the first time. The "Tianhe-2" again ranked the first in the world. In other words, whether based on Paper A, which examines the computational speed, or based on Paper B, which examines the application performance, the "Tianhe-2" was tested and proved to be the champion!

Therefore, the "Tianhe-2" is a truly a world champion and a "double-standard" champion!

Perhaps no amount of proof will ever quieten the critics. They will always find some aspect to criticise. The best response is to ignore them. Their captiousness only helps China's supercomputer technology to develop even faster. The Tianhe researchers have never considered the championship so highly. Just as the award-winning representative Lu Yuxi said, "Despite the fact that the 'Tianhe-2' has won several consecutive championships, there is still a long way for Chinese supercomputers to go to be first in the world. The United States is still in a dominant position in the supercomputer field."

V. Pillars of the Country

With mankind continuously expanding and deepening its knowledge, especially with the emergence of big science and megaprojects in modern times, there have been growing difficulties in supporting the increasingly huge scientific structure through traditional approaches of exploration. Instead, mathematical thinking and approaches summarized as high-performance computing are needed. In response, electronic computers were built in the middle of the 20th century. After rapid development over 50 years, high-performance computing is currently an integral part of the innovation platform for advanced technologies.

What can high-performance computing, i.e., supercomputers,

actually compute?

It is safe to say that there are no other disciplines of our times used in such a wide, deep and cutting-edge way in scientific research as high-performance computing. As Liu Guangming, Director of the National Supercomputer Center in Tianjin, has said: "A supercomputer can compute data about the sky, the Earth, the people, the past, the present, the future...oil can be found quickly and precisely when doing a CT on the Earth with a supercomputer. The secret of life can be decoded if a supercomputer is used to analyse human genes. When a supercomputer is used to design a wind tunnel, the aircraft can fly faster and higher with less fuel."

Currently, there are challenges from nearly all aspects of science which need to be addressed by supercomputers. Some of them are listed below.

Manufacturing of vehicles: Supercomputers can be used to identify and improve the aerodynamic/hydrodynamic structure, fuel consumption and impact strength of vehicles such as cars, airplanes and ships, helping to reduce noise and enhance comfort for passengers.

Weather forecasting: A model built by a supercomputer can be used to predict changes in the climate and prevent and alleviate damage caused by climate change.

Biological information: Supercomputers will help mankind find revolutionary solutions for treating diseases, with approaches from data-intensive research on genetics to simulation of cellular networks and large-scale system modelling.

Seismic monitoring: Simulation of earthquakes by supercomputers will help mankind explore new ways to predict such disasters and reduce losses of life and property through early warnings.

Geosciences: Geophysics requires significant data processing and simulation. Therefore, supercomputers will potentially bring huge economic benefits to activities such as oil exploration.

Astrophysics: Simulation by supercomputers is the foundation for the research on astrophysics. Modelling and theoretical experiments on the evolution of celestial bodies can be done with supercomputers, simulating and accelerating the passage of time.

Public health: Supercomputers are capable of simulating events which impair the health and safety of mankind and proposing measures and plans in response to possible major incidents of pollution and disasters.

Materials science: Substances and reactions with huge economic value can be discovered through intensive computing and the simulation of substances and energy by supercomputers.

Research on anthropological/organizational systems: Supercomputers can be used to deeply analyse macroeconomics and social dynamics so as to find the laws by which mankind carries out its activities and by which a society develops.

...

With such tremendous power in boosting and accelerating scientific advances and social development enjoyed by supercomputers, scientists embarked on the promotion and application of high-performance computing more than 30 years ago.

In 1983, only several years after the advent of the "Cray-1", the world's first supercomputer, Larry Smarr, a professor from the University of Illinois, proposed that the National Science Foundation should launch a centre for supercomputing. This proposal was supported by a large number of scientists. Soon, the first two centres for supercomputing in the world were established in the United States. Since then, developed countries such as Germany, France, the United Kingdom and Japan and developing countries such as China and Brazil have all followed suit, leading to the flourishing of supercomputing all across the globe. Related facilities were built one after another, including Oak Ridge National Laboratory in the United States, equipped with well-known supercomputers such as the "Jaguar" and the "Titan", Lawrence Livermore National Laboratory in the United States with supercomputers named "Sequoia" and "Blue Gene", the Riken Advanced Institute for Computational Science in Japan with the "K computer", Argonne National Laboratory in the United States with the supercomputer named "Mira", Forschungszentrum Jülich or Forschungszentrum Juelich research centre in Germany with the "Juqueen", Leibniz Supercomputing Centre with the "SuperMUC", the Texas Advanced Computing Center with the "Stampede", the Chinese National Supercomputing Center of Tianjin with the "Tianhe-1", CINECA in Italy with the "Fermi", the "IBM power775" used by the United States Defense Advanced Research Projects Agency and the Chinese National Supercomputer Center in Guangzhou with the "Tianhe-2".

Together with the establishment of these high-profile supercomputing centres, high-performance computing has become an innovation platform for various disciplines and an important symbol of a country's competitiveness in science and technology.

Moreover, with the seeding and sprouting of high-performance computing in different academic fields and its blossoming in various areas of our daily lives, mankind is increasingly experiencing every benefit from the cultivation of these technologies.

For example, in terms of food, scientists are using supercomputers to perform research on genetic engineering, with subjects like rice, corn and pigs. Let us take pigs as an example. Since this animal might have certain genetic defects, it grows slowly and is susceptible to diseases, with more fat meat and less lean meat. Therefore, some dishonest farmers feed their pigs with prohibited drugs such as clenbuterol hydrochloride under the purpose of improving quality of the meat, thus posing a danger to the human body. However, supercomputers can be used to find the genetic defects so that effective rectification can be done to the extent that the pig would have fast and healthy growth, benefiting mankind.

Another example is disease treatment. It is widely known that new viruses threatening human bodies are arising one after another with strange and versatile features. If we think of a new virus as a target, the course by which we develop medicine against it would be filled with target practice. Researchers have to perform experiments repeatedly in order to find the drug molecules which are capable of fighting against the virus. It is like looking for a needle in a bundle of hay and takes from three or five years to over 10 years. With supercomputers, hundreds of thousands of pairings can be done within a very short period of time, accelerating the targeting process, with effective medicines being found within as short a period as several months. Furthermore, when a patient with cancer needs chemotherapy and radiotherapy, the cancer genes must be identified beforehand. Previously, this would take one or two months. Today, thanks to supercomputers, everything is figured out within just several minutes. Two months are vitally important for a patient in critical condition.

There are also examples related to traveling. It is the habit of nearly everyone to check the weather before traveling. Whether the weather forecast can be made accurately, promptly and well in advance hinges on whether there is enough observational data and sufficient computing power. A supercomputer can effectively break these two bottlenecks. Within one day, it can complete computing which would normally take several years and even several decades in the past. In addition, a supercomputer can also be used to predict natural disasters such as earthquakes and tsunamis.

In terms of entertainment, everyone was amazed by the special effects in the movie *Avatar*. The rendering of effects in this movie was performed by a supercomputer. The Chinese movie *The Lost*

Bladesman, the new version of the TV series *Journey to the West* and various American movies like *Resident Evil* are also masterpieces of the "Tianhe" supercomputer.

...

Currently, high-performance computing has penetrated into every industry and millions of households. All the time, people are sharing the benefits brought by supercomputers in terms of clothing, food, housing, transportation and entertainment.

Supercomputing is definitely one of the pillars of the country of our era.

VI. Superhero

China's supercomputer application started with the birth of its first supercomputer, the "Yinhe-1". In the 1980s, the "Yinhe-1" was successfully applied in different areas including aerospace, earthquake prevention, geological exploration, weather forecast and applied physics, fostering the first users of China's computing. In the 1990s, with the success in the research and development of the "Yinhe" series, the "Sugon" series, the "Shenwei" series and the "Shenteng" series, both application areas and users were greatly expanded. However, in general, supercomputing in China was still in oversupply during this period, with computing hardware lying idle. There was even a phenomenon of paying for users at that time.

At the turn of the century, the Shanghai Supercomputer Center and the Supercomputing Center of the Chinese Academy of Sciences were established and officially opened to the public, which launched a new era of supercomputing centre construction and supercomputing applications in China.

In 2002, the 16th National Party Congress proposed the strategic goal of building an innovative country and the following National Science and Technology Congress launched its National Guideline on the Medium- and Long-Term Program for Science and Technology Development (2006–2020). As an important symbol of the national scientific and technological level, the supercomputer was listed as a key development project, of which the promotion and application was a priority. Provincial and municipal governments began to build supercomputing centres and the construction of these centres formally entered a phase of rapid development. Supercomputing centres no longer needed to pay for users, and instead witnessed users waiting for machines.

By 2004, the application of China's several supercomputing centres had been fully saturated. Afterwards, China's demand for supercomputing grew at a rate of 10–20% every year, and now the rate reaches 20–30%.

Currently, China has built five national-level supercomputing centres which are NSCC-TJ, the National Supercomputer Center in Jinan (NSCC-JN), the National Supercomputer Center in Changsha (NSCC-CS), NSCC-SZ and NSCC-GZ. Others are located in Chengdu, Hefei, Shanghai and elsewhere.

Established by the NUDT and Binhai New District in Tianjin, NSCC-TJ is the earliest and the largest national-level supercomputing centre in China. It is equipped with the "Tianhe-1" (Phase 2 System) which consists of 140 cabinets and which ranked first in the international TOP500 in November 2010. The centre also owns three sets of supercomputing systems which are the "Tianhe Tianteng", with an operating speed of 100 teraflops, the "Tianhe Tianxiang" with 128 Intel-EX5675CPUs and the "Tianhe Tianchi" with 96 CPUs.

With high-end, systematic and well-developed configuration, NSCC-TJ has become one of China's best centres in terms of its computing capabilities. It is just like a ploughshare, developing fertile soil for China's high-tech field.

Modern people cannot live without oil, just like they cannot live without water. But where does oil come from? When it comes to this question, people will be reminded of a mental image in which several geological explorers in red work clothes and red security helmets, with heavy equipment on their backs, are tapping rocks by the rivers and lakes using a small hammer. Fast-growing computer technology has already eliminated this kind of primitive method for oil exploration. People now are able to restore the geological structures of a certain region and identify the exact location and reserves of oil and natural gas by applying supercomputers to perform scientific calculations on three-dimensional seismic data and build large-scale three-dimensional geological models. The faster the operating speed, the faster and more accurate the exploration will be.

Collaborating with relevant domestic research institutions, NSCC-TJ has established the "Tianhe Petroleum Geophysical Prospecting Center" and built a perfect seismic data processing and geophysical exploration information management platform especially for CNPC, Sinopec and CNOOC. It also greatly facilitates the processing, development and application of petroleum geophysical data and has successfully designed a three-dimensional oil data displacement software with completely independent intellectual property rights, reversing the passive situation in this field and promoting industrial technological progress and collaborative innovation.

Based on these technological advantages of "Tianhe-1", the Bureau of Geophysical Prospecting INC., CNPC has gradually finished several thousands of square kilometres of continuous and high-frequency seismic big data projects.

What was our planet like hundreds of thousands of years ago? And how did it become the way it is today? To solve these mysteries of the Earth, it is far from enough just to be like Copernicus, Bruno, and other ancestors who only depended on their wisdom, eyes and astronomical telescopes. Now we must rely on supercomputers for numerical simulations to research various physical, chemical and biological changing processes in a comprehensive and systematic way. The "Tianhe-1" is outstanding in the research of this field.

The "Tianhe-1" has become the research and development simulation platform for the Institute of Atmospheric Physics of the Chinese Academy of Sciences, the State Oceanic Administration and the China Meteorological Administration. By simulating global climate change and changes in the marine environment, it forms a certain understanding of the living environment of human beings in the future and provides scientific data to ensure the sustainable development of human society.

The "LICOM Study on High-resolution Ocean Circulation Model" conducted by the "Tianhe-1" is the first in which China used its own vortex-resolution ocean model to complete long-term numerical simulation tests and to obtain accurate and effective test data.

Thanks to the "Tianhe-1", a heat flow simulation of the Earth's outer nucleus, completed by the Institute of Software of the Chinese Academy of Sciences, has realized an ultra-large-scale numerical simulation of 60 billion unknowns for the first time in the world.

"Compared with the supercomputing systems such as IBM's 'Blue Gene' that world powers take pride in, the computing results of the 'Tianhe-1' can totally match them and be even better!", praised the scientists from the Chinese Academy of Sciences.

Supporting independent intellectual property drug development is one of the important missions of NSCC-TJ.

Among the contributions that the "Tianhe-1" has made to human health, one of its important users, Tianjin International Joint Academy of Biomedicine (TJAB) is a good example. Scientists from the Academy said, "In the past, it took us one billion United States dollars, 10 years and 100,000 iterations of compound screening to develop a new drug. However, thanks to the 'Tianhe-1', we can now finish the screening process within one week which would have been completed in a year in the past. It greatly shortens the research and development cycle, reduces costs and provides new ideas for the development of new drugs. The application of high-performance computing in drug development is a revolution in the pharmaceutical industry."

Scientists from the Shanghai Institute of Materia Medica of the Chinese Academy of Sciences had the same feeling, saying, "By combining computing simulation and drug testing on the 'Tianhe-1', we confirmed a completely new drug site. Direct drug design in NSCC-TJ, without any chemical process, has achieved a drug that has a significant effect on a certain disease. We have also successfully carried out the fundamental research on new relevant GPCR (G Protein-Coupled Receptors) drug targets for major cardiovascular disease."

The successful application of the "Tianhe-1" in drug development is sufficient to rebuild confidence for patients of AIDS, epilepsy, diabetes and other diseases in overcoming chronic illness.

Aerospace is the focus area that world powers are ready to put ahead of their own rivalries. The "Tianhe-1" is a major platform for designing aerodynamics simulations and developing new engines for aerospace vehicles in China. It has currently won a massive number of significant users such as the Institute of Aerospace A, Peking University College of Engineering, the National Lab of

Computational Fluid Dynamics, the Beijing Aeronautical Science & Technology Research Institute, the Chinese Flight Test Establishment and the Institute of Mechanics of the Chinese Academy of Sciences. It has also already completed the important subject research on the "Simulation of Pneumatic Machine of Large Aircraft" and "Virtual Test Environment of Large Aircraft based on Models".

Atmospheric haze has been an increasingly problematic and dangerous problem. In order to figure out the formation areas and process of haze and the relevant physical, chemical and biological factors to build related models and make precise predictions, Institutes like the Chinese Academy of Meteorological Sciences, the National Meteorological Center, Tsinghua University and NSCC-TJ work together to establish and improve the digital simulation model with the 10–20% computing resource of the "Tianhe-1". Currently, critical progress has been made in the research, and precise predictions will probably be made about the haze three to five years later. Relevant evaluated data are to be provided for China to work out regional development projects.

A "Wisdom City" is not only a buzzword that people talk about frequently, but also a receding vision. Wisdom City is defined to realize the merge of IOT infrastructure, cloud computing and geospatial arrangements, to perceive a city's operation comprehensively, to respond to city demands intelligently and to manage city life smartly. The hundreds of elements like public transport, public health, public information and public energy resources all require the support of supercomputing technology.

NSCC-TJ plays a positive role in the establishment of Wisdom Cities by conducting pilot construction of the national Wisdom Cities in Tianjin and Zhengzhou, and has accumulated valuable technology experience. In the future, they will also strive to make crucial technological breakthroughs of big data, the integration of high-performance computing and big data and the integration of cloud computing and big data. They will build supercomputing platforms for new industries in oil and gas exploration in northern China, biological information, environmental applications, transportation uses, new energy resources and decision-making of the government, social security and science management, to provide technological support for the big data of the "Wisdom Binhai New District".

In field of mega engineering, the "Tianhe-1" has also made many commendable contributions.

The "Tianhe-1" has solved the bottleneck problems of long cycle time and low accuracy in the whole process of large and complex parts, saved funds in research and development and increased international competitiveness of the enterprise by greatly reducing the cycle time of research and development and improving product quality.

Tianjin Motor Dies Company Limited has been the professional partner of international super-class enterprises like the General Motors Company (GM) and Benz, using the international advanced level technology of six-million-unit simulation enabled by the "Tianhe-1".

The School of Chemical Engineering of Tianjin University has completed the three-dimensional process simulation of heat recovery in a coke oven on the "Tianhe-1" for the first time.

First Auto Works in Tianjin (FAW TJ) has applied the "Tianhe-1" to carry out security design and simulation; thus the three days of vehicle crash simulations has been reduced to only two hours.

...

Furthermore, the "Tianhe-1" systems allocated to NSCC-CS have greatly improved the ability of high-performance computing in central China and have noticeably enhanced the public capability of weather forecasting, disaster prevention and environmental protection and the service of universities and research institutes since being put into operation in Hunan University in 2011. In particular, they have provided computing platforms for industries like equipment manufacturing, metallurgy of iron and steel, automotive production, biomedicine and animation design.

The application achievements of the “Tianhe-1” are like stars in the galaxy, and it is difficult to keep count of them. Sets of data are prepared to prove that.

Since NSCC-TJ opened, the “Tianhe-1” has provided more than 500 key users with high-performance computing services, including some high-end users like CNPC , Sinopec, CNOOC, BGI, the Tianjin Institute of Pharmaceutical Research, CSIC, CFHI, Tianjin Motor Dies Company Limited, the Chinese Academy of Sciences, National Geomatics Center of China, the central forecasting office of the State Oceanic Administration, China Meteorological Administration, Academy of Military Medical Sciences, NAOC, Tianjin University, Peking University and the University of Science and Technology of China.

On the “Tianhe-1”, there are over 1,000 application programs running at the same time, and several hundred corresponding innovation programs, as well.

The “Tianhe-1” supports more than 50 national “863” and

"973" projects, nearly 400 projects of the National Natural Science Foundation of China and more than 20 other significant projects.

The "Tianhe-1" has a high usage level, above 80% on average, and is overburdened compared with others in the field. Almost every user with demands for supercomputing in China has used the "Tianhe-1".

The moment of the birth and championship glory of the "Tianhe-2" was exactly the time at which the CPC Central Committee with General Secretary Xi Jinping called for the "Chinese Dream". It was also the crucial period in which the Chinese economic development pattern was transformed from "Made in China" to "Created in China". China's aspirations and the demands for scientific innovation have functioned as two powerful magnetic fields attracting users to the "Tianhe-2". There are significant uses such as the large-scale genome-wide analysis of millions of people, high-throughput fake drug screening, the structural design of large equipment and large-scale simulation of dark matter in cosmology flooding into NSCC-GZ.

The only thing they can do is to debug the system while it is in service. More than 80 new users carry out over 100 "863" and "973" projects in applied computing in the course of six months. There were six new large-scale applied field applications over a million of compute cores merely in April of 2014. Now, the “Tianhe-2” has nearly 200 users.

These applied studies have achieved great success in many fields like the design of large aircrafts, high resolution Earth observation, DNA sequencing, biomedicine, establishing Wisdom City, e-commerce, cloud computing, information services, etc., and have ushered in noticeable economic and social benefits as well as many awards and world-firsts.

The high degree of aerodynamic computation of the flow field of the COMAC C919 airliner on the “Tianhe-2” was the first to realize a high-precision large-scale numerical simulation of complex configuration of an aircraft worldwide, and provided crucial scientific data for model developments.

The ultra-high-resolution global medium-range numerical weather prediction, with the help of “Tianhe-2”, has achieved the highest resolution in the world, which enables China to lead the world in this field.

The Shanghai Institute of Materia Medica at the Chinese Academy of Sciences conducted affinity assessment of 750,000 small molecule compounds and completed over 600 drug tests and evaluation of in vitro and in vivo activity on the "Tianhe-2", which offered new approaches to fight stubborn diseases like malignant tumours, hepatitis B, diabetes, etc.

...

BGI, the main producer of China's biological data, is also a typical user of "Tianhe" supercomputers. After the second phase of the system of the "Tianhe-1" settled down in Tianjin, BGI used approximately half of its computing. BGI was again one of the significant users of the "Tianhe-2" after its birth.

Why is BGI is so fond of the "Tianhe" series? The chairman of BGI answered, saying, "We feel sorry for the hardships the supercomputer in China has been through. We at BGI also experienced the same twists and turns." Today, BGI, with the most powerful life big data generating capacity worldwide, and the "Tianhe" series, with the most powerful computing capability in the world, integrate closely to conduct biological super data computing, bringing fundamental reforms to bio-breeding and medical education. Gene technology is about to benefit mankind and usher in an emerging bio-economy era.

With the commencement of the "Tianhe-2", NSCC-GZ that has integrated high-performance computing, mass data processing and information management services (MIS), will be widely applied in high-technology industries, modern service industries, digital city and scientific research and innovation etc., a powerful engine of industry transformation in the southern part or even the whole country.

NSCC-GZ planned to speed up the computing to more than 100 petaflops by 2015, becoming the world's first supercomputing centre.

"Tianhe" supercomputers have also completed over 10 international or regional projects beyond China's borders, becoming an ambassador for China's international cooperation in this field.

The NVIDIA Corporation offered to sign an agreement establishing the top joint laboratory in the world together with NSCC-TJ. The agreement indicated that the more advanced GPU technologies of the NVIDIA Corporation would be applied to enhance the supercomputing capacity and usage efficiency to contribute to the intimate integration of cloud computing, IOT and the Wisdom City concept with the unique communication processing technology of the "Tianhe-1". As the scientists predicted, the establishment of a joint laboratory would save hundreds of millions of RMB and create substantial values for modern service-oriented enterprises in Tianjin Binhai New District, boosting the international competitiveness of the area.

After 2011, NSCC-TJ experienced extensive exchanges and cooperation with other supercomputing centres and high-performance computing institutes in America, APAC countries and Europe. The research institutes such as Oak Ridge National Laboratory, Lawrence Livermore National Laboratory, Lawrence Berkeley National Laboratory, Argonne National Laboratory, Northwest Pacific National Laboratory and the German Forschungszentrum Juelich research centre sent delegations to NSCC-TJ and conducted a number of collaborative studies. In 2012, the Ministry of Science and Technology designated NSCC-TJ as "The International Science and Technology Cooperation Demonstration Base of Supercomputing Technology Innovation and Research and Development."

After assembling the "Tianhe-2", NSCC-GZ signed a strategic memorandum of cooperation with the United Kingdom National Supercomputing Centre, opening the doors to serve foreign research institutions. The Technical University of Munich, Germany, used the "Tianhe-2" to simulate the 1992 earthquake in Landers, California, whereas the Georgia Institute of Technology in the United States carried out research on an anti-HIV drug action mechanism. The research results obtained by the Technical University of Munich are shortlisted for the "Gordon Prize".

The 21st century, for China's supercomputer applications, can be described as the spring of the Earth, in which everything is thriving and prosperous. At the same time when the "Tianhe" supercomputer applications were in harvest season, other supercomputers and supercomputing centres with domestic models also provided a powerful boost for the country to transform from "Made in China" to "Created in China".

Founded in 2000, Shanghai Supercomputing Center with the deployment of the "Sugon" supercomputers, Fluent for CFD, CFX, Ansys for structural analysis, Nastran, Ls-Dyna for nonlinear dynamic analysis, Pam-crash and other large-scale applications, is China's earliest constructed, shared resource high-end computing platform. The Center, adhered to the principle of service "based in Shanghai, facing the country", not only provided a strong support for the high-performance computing of scientific research, but also provided quality services for engineering computing in a number of industries, especially manufacturing, and has been successful and widely used in the aerospace, aviation, shipbuilding, nuclear, metallurgy and municipal engineering industries, as well as other fields.

At present, the Shanghai Supercomputer Center is integrating the fragmented business software based on business needs, constructing professional integrated design and analysis platforms of automobiles, aircraft and ships, to provide users with more powerful and more professional computing services.

The Supercomputing Center of the Chinese Academy of Sciences, which has the largest computing platform in China, is also a supercomputing centre for the whole society, providing users with the research, implementation and application services of parallel processing, and providing solutions for research on large-scale complex technologies and commercial applications. So far, more than 300 users in the Center have been on the computer for 30 million core-hours, which has played a key role in helping a series of national “863” “973” projects to conquer core and key technologies.

The “Earth System Simulator” supercomputer, configured at HPC, Tsinghua University, achieves an operating speed of 172 teraflops, the fastest-running supercomputer in Chinese universities. The system has been tasked with the calculation of the climate simulation, prediction and evaluation trials of the Fifth Assessment Report (IPCC-AR5) of the Intergovernmental Panel on Climate Change (IPCC) in the United Nations. In particular, the “Earth System Simulator Research Project” jointly carried out by Tsinghua University and the Inspur Group in this system will provide an important basis for China to take countermeasures related to environmental change, environmental protection and scientific utilization of resources, and has important practical guiding significance for advancing the Chinese Earth system scientific research and China's economic and social development.

...

Demand in key areas of national development and defence security has always been the primary driving force behind the development of supercomputers and the most important application area. Exactly because of its success in using the supercomputer platform, the United States supported and promoted a series of scientific and technological innovation, built the world's largest economy, established the most advanced, system-wide, powerful defence military equipment system, and dominated the world.

Thankfully, modern China is fast running to modernity with unstoppable momentum. China's success in supercomputing is one of the striking examples of this.